# Avian (Poultry) Production

# Avian (Poultry) Production

*Editor*

Carin Fisker

scitus
academics

**Avian (Poultry) Production**
Edited by **Carin Fisker**

ISBN: 978-1-68117-416-7
Library of Congress Control Number: 2016937427


SCITUS Academics LLC,
www.scitusacademics.com
Box No. 4766, 616 Corporate Way,
Suite 2, Valley Cottage,
NY 10989



Printed in United States of America on Acid Free Paper

# Preface

Poultry has a crucial place worldwide as the eggs and chicken meat are important and rich sources of protein, vitamins and minerals. Poultry provides rich organic manure and is an important source of income and employment to millions of farmers and other persons engaged in allied activities in the poultry industry. Chickens and turkeys that are commercially raised today are the result of many years of selection and breeding of two or more breeds to continually improve productivity. Poultry farming is the raising of domesticated birds such as chickens, ducks, turkeys and geese for the purpose of farming meat or eggs for food. Poultry are farmed in great numbers with chickens being the most numerous. More than 50 billion chickens are raised annually as a source of food, for both their meat and their eggs. The chicken industry is made up of meat-producing chickens and egg-producing chickens. The modern meat-type chicken is derived from the deep-breasted Cornish breed and has been selected with an emphasis on producing meat rather than eggs. Further processing of meat chickens has led to many new products geared to the convenience-minded consumer, such as boneless, skinless breasts. The turkey industry has developed strains of birds with an emphasis on meat production and breast size, as the white meat has a higher market value in the United States. Many pure breeds of poultry are raised for exhibition, hobby and/or breed preservation purposes. There is lots to discover with the comprehensive poultry production. We all know that in order to get the best out of birds, poultry keepers need to be up to date with the latest ideas and techniques. The book Avian (Poultry) Production discusses about the anatomy and physiology of birds, building them the best housing, breeding, egg production and also about preventing parasites and diseases.

# Table of Contents

# CHAPTER 1

# Artificial Insemination in Poultry

**M.R. Bakst[1] and J.S. Dymond[1]**

[1] Animal Biosciences and Biotechnology Laboratory Beltsville Area, Agricultural Research Service, U.S. Department of Agriculture Beltsville, Maryland, USA

## 1. INTRODUCTION

Artificial insemination (AI) is the manual transfer of semen into the female's vagina. Basically it is a two step procedure: first, collecting semen from the male [1]; and second, inseminating the semen into the female [2]. In poultry, depending on the objectives and goals of the farm or laboratory, there may be intervening steps such as semen dilution, storage, and evaluation.

Artificial insemination is practiced extensively with commercial turkeys. This is primarily the result of selective breeding for a heavier and broader-breasted commercial turkey and the consequent inability of toms to consistently transfer semen to the hen at copulation. The broiler industry has not adapted AI to the extent of the turkey industry but it is occasionally used in pedigree lines and in regions where labor is relatively cheap.

To grasp the magnitude of AI in the turkey industry compared to that of livestock, a hypothetical flock of 500 breeder hens inseminated with 100 μL of diluted semen (1:1) twice the week before the onset of egg production and once weekly thereafter for the 24 wk of egg production would entail 13,000 inseminations using 650 mL of semen. It should be apparent with these numbers, semen collection and hen inseminations are labor intensive as each male and female must be handled each week.

Looking back over the use of AI in the turkey industry one can safely say that in the 1960s, weekly inseminations were based on semen volume per dose using undiluted semen. In the 1970s and early 1980s, breeder farms began to dilute semen and inseminate a known number of sperm per dose. In the mid-1980s through the 1990s, hens were initially inseminated a week before the onset of lay and inseminations were performed with a known number of 'viable' sperm. Currently, while inseminating before the onset of egg production remains widely practiced, most companies, but not all [3], have gone back to inseminating a known volume of semen or number of sperm per dose, in the 1970s and 1980s.

In the following chapter we will review the basics of AI and fertility evaluation in poultry. To better appreciate the biological basis of these techniques, an overview of the reproductive biology of poultry is provided. Detailed descriptions of techniques for the collection, evaluation, dilution, and storage of poultry semen are available in a recent publication by Bakst and Long [4]. Earlier comprehensive reviews include Lake and Stewart [5], Bakst and Wishart [6], and Bakst and Cecil [7].

## 2. REPRODUCTIVE BIOLOGY OF POULTRY

This section will introduce to some and review for others the strategy of avian reproduction with emphasis on the hen. For more comprehensive reviews on reproduction in the avian male and female see Jamieson [8].

## 2.1. Overview

The goal of AI is to produce a succession of fertilized eggs between successive inseminations. To accomplish this, weekly inseminations must replenish the sperm population in the uterovaginal junction (UVJ) sperm storage tubules (SSTs). Birds do not have an estrous cycle that synchronizes copulation with ovulation. Alternatively, about 7-10 days before their first ovulation, hens mate, sperm ascend the vagina and then enter the SSTs. At the onset of egg production, individual sperm are slowly released from the SSTs, transported to the anterior end of the oviduct, and interact with the surface of the ovum (see [9-10] for recent reviews). Whether fertilized or not, over the next 24-26 hr the ovum is transported though the oviduct, accruing the outer perivitelline layer (PL) in the infundibulum, the albumen in the magnum, the shell membrane in the isthmus, and the hard shell in the uterus (also referred to as the shell gland) before oviposition. If fertilized, the blastoderm in the first laid egg consists of 40,000-60,000 cells in the turkey and 80,000-100,000 cells in the chicken.

Ovary: In the hen only the left ovary and oviduct become functional organs. About 2-3 wk before the onset of lay, small (less than 1 mm in diameter) white-yolk follicles begin to accumulate yellow yolk with some being recruited into a hierarchy of maturing yellow-yolk follicles (Figure 1). At the time of ovulation, the largest follicle, designated as F1, is ovulated. About 17 days were necessary for the 1 mm diameter white yolk follicle to mature to a pre-ovulatory 40 mm diameter yellow yolk follicle [11]. After the F1 follicle is ovulated, the next largest follicle, formerly designated F2, becomes the F1 follicle and will ovulate at the beginning of the next daily "ovulatory cycle" in 24-26 hr.

The follicular sheath surrounding the maturing oocyte consists of histologically distinct concentric layers of cells: the outer serosa (germinal epithelium); the theca externa, which forms the greatest portion of the follicle wall, provides structural support to the follicle and has steriodogenic cells; the theca interna, a highly vascularized layer, which like the theca externa has steroid-producing cells (both thecal layers synthesize androgens and estrogens); and, the granulosa cell layer, enveloping the oocyte, which is responsible for progesterone secretion and the synthesis of the inner PL. The inner PL is homologous to the mammalian zona pellucida and is a fibrous reticulum about 2 μm thick. At ovulation, only the inner PL envelops the ovum. While there is no corpus luteum formation in birds, the thecal layers and the granulosa of the post-ovulatory follicle (POF) produce prostaglandins [12] and progesterone, respectively [13-16] then regress over the next 72 hr. The POF has a pocket like appearance after ovulation (Figure 1). On the surface of the inner PL overlying the germinal disc (GD), which is a 3.5 mm diameter disc of white yolk containing the haploid pronucleus and associated organelles, are sperm receptors. Sperm bind to the receptors overlying the GD, hydrolyze a path through the inner PL, and are incorporated into the ovum. Polyspermy is normal in birds but only one sperm in apposition to the female pronucleus undergoes nuclear decondensation and initiates syngamy, the reconstitution of the diploid number of chromosomes.

## 2.2. Oviduct

The mature oviduct consists of five anatomically and functionally distinct segments (Figures 1 and 2): the infundibulum, which secretes an albumen-like product that forms the outer PL and prevents pathological polyspermy; the magnum, responsible for deposition of the albumen proteins; the isthmus, which forms the shell membrane; the uterus (also referred to as the shell gland), a pocket-like structure that elaborates the hard-shell; and, the vagina, which is a conduit between the uterus and cloaca for the egg-mass at oviposition and is responsible for sperm selection and storage following semen transfer. Interestingly, when the vagina and uterus are excised and fixed *in toto* and the connective tissue surrounding the vagina subsequently removed, the vagina appears as a coiled segment (Figure 3) [10]. This anatomy explains the resistance one feels when performing a vaginal insemination with a straw regardless of the presence or absence of an egg mass in the uterus. If inseminating a hen within 30 min after oviposition, the connective tissue around the vagina and the smooth muscle composing the vaginal wall are flaccid. Venting (exteriorizing the vagina for placement of the inseminating straw) at this time may induce a partial prolapse leading to a deep insemination (closer to the UVJ) and the forfeiture of sperm selection by the vagina. Such deep inseminations are associated with high embryo mortality, possibly due to pathological polyspermy.

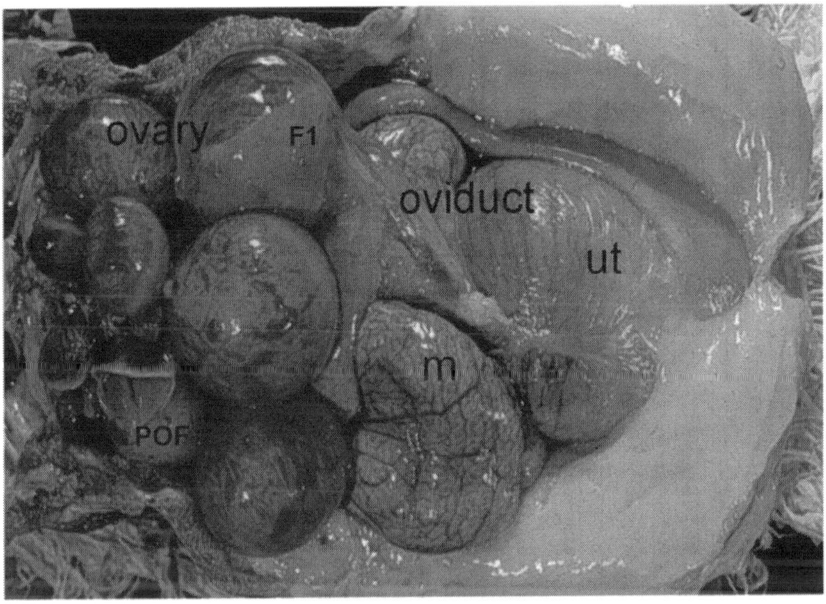


**Figure 1.** The ovary and oviduct of a turkey hen in egg production occupy much of the abdominal cavity. The ovarian follicular hierarchy consisting of ovarian follicles at various stages of develop (7 maturing follicles visible in this photograph) is observed. The largest follicle, F1 follicle is the next to ovulate. The ovum ovulated about 10 hr earlier has accrued albumen in the magnum (m), a shell membrane in the isthmus, and is observed in the uterus (ut) undergoing shell formation. Its post-ovulatory follicular sheath (POF) appears as an open pocket. The vagina (distal to the uterus and not visible) is embedded in connective tissue and enveloped by the abdominal fat pad.

The surface mucosa of each segment of the oviduct is lined with parallel, gently spiraling folds along the longitudinal axis. The surface epithelium lining the luminal mucosa contains varying proportions of secretory and ciliated cells. All segments except the fimbriated region of the infundibulum and the vagina possess sub-epithelial tubular glands that secrete components used in egg formation [17]. However, the anterior 2-

3 cm of the vagina, an area referred to as the UVJ (Figure 3), contains the SSTs, the primary sites of sperm storage [10] (Figure 4).

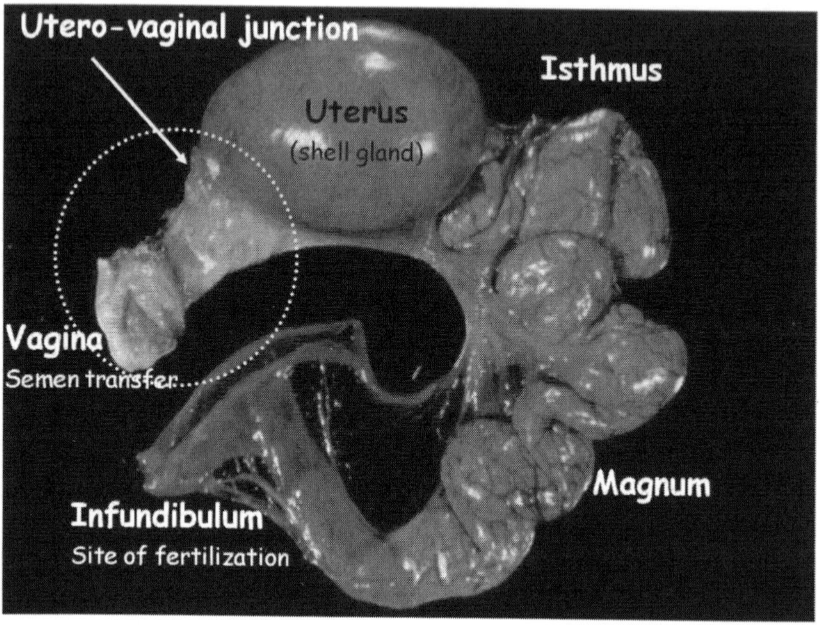


**Figure 2.** The segments of the turkey oviduct with a hard-shelled egg in the uterus are observed. Sperm transferred into the vagina undergo an intense selection process before reaching the sperm storage tubules (SSTs) localized in the utero-vaginal junction. Sperm are slowly released from the SSTs and ascend to the infundibulum, the site of fertilization. In this photograph, the vagina is enveloped by connective tissue.

At ovulation, the ovum is grasped by the fimbriated region of the infundibulum and, if sperm are present, the ovum may be fertilized within a 10-15 min interval [18]. Thereafter, infundibular secretions accrue around the ovum, forming the outer PL, which acts as a barrier to further sperm penetration. Birkhead [19] observed that the number of sperm trapped in the outer PL is positively correlated with the size of the ovum and is likewise correlated with the number of sperm that have penetrated the inner PL. Interestingly, the sperm trapped in the outer PL retain an intact acrosome [20-21]. If fertilized, the first cleavage furrow in the GD appears 7-8 hr post-ovulation, while the egg-mass is in the isthmus.

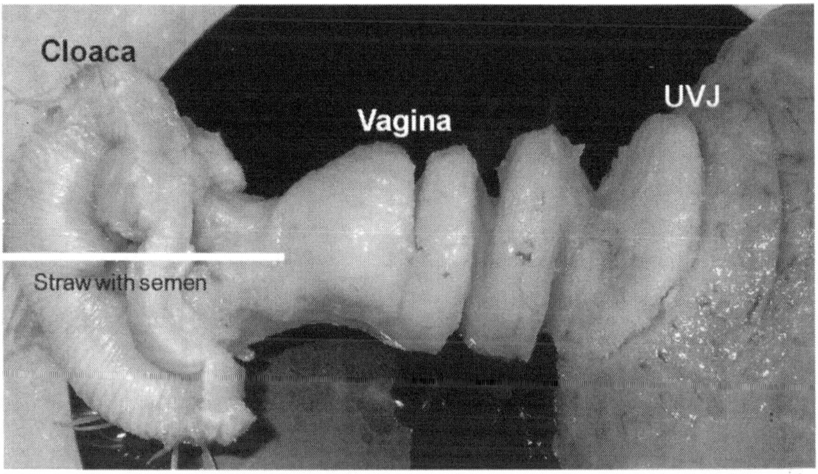


**Figure 3.** Following fixation in neutral-buffered formalin and the removal of the surrounding connective tissue, the coiled morphology of the turkey vagina is revealed. When inseminating a hen, one should insert the straw with the semen until resistance is felt, then release the semen as the straw is withdrawn. As observed here the resistance is due to the coiled vaginal and not an egg mass in the uterus.

## 2.3. Oviductal Sperm Selection, Transport, and Storage

Following deposition in the oviduct, sperm are transported to UVJ by a combination of their intrinsic motility and cilia beat activity [9-10, 22-23]. Within the SST lumen, sperm are either widely spaced or oriented parallel with their heads toward the distal end of the SST (Figure 4). Sperm are apposed to, but not directly contacting the apical microvilli of the SST epithelial cells. This spatial relationship may facilitate lipid transfer between the resident sperm and the SST epithelial cells [24-25]. Interestingly, alkaline phosphatase, known to play a role in lipid transfer, has been histochemically localized in the apical region of the SST epithelium [26].

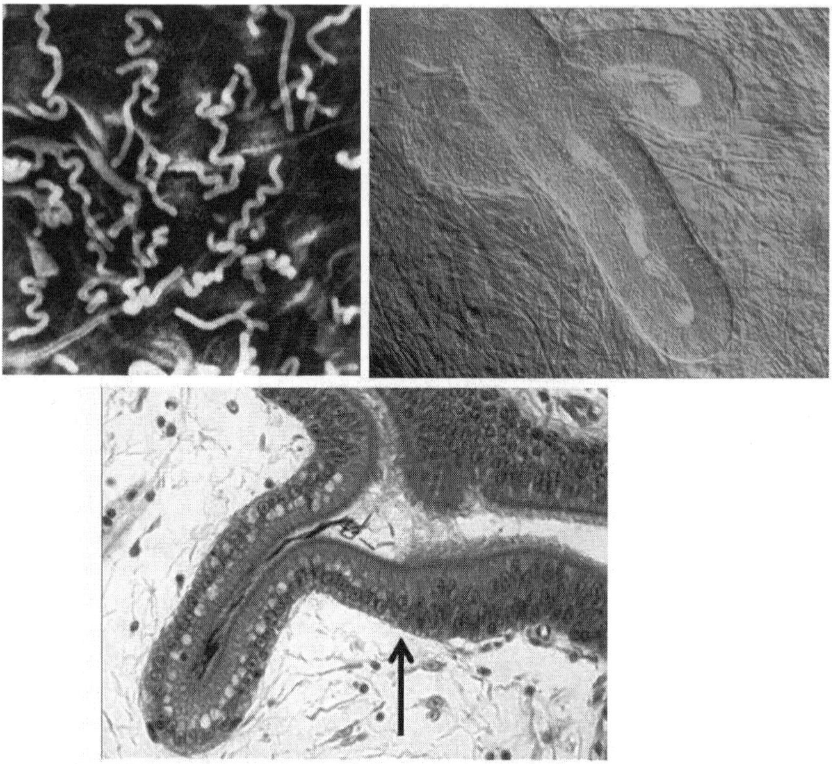

**Figure 4.** Three views of the turkey's sperm-storage tubules (SSTs) are observed. The left panel is a stereoscope image showing the pleomorphic appearance of the SSTs. The length of the SST can be as long as 300μm. In the right panel a hen was inseminated with sperm stained with Hoechst 33342, a nuclear fluorescent dye, the UVJ mucosa containing SSTs was isolated, and an unfixed squash preparation was observed by dual interference contrast and fluorescence microscopy. Sperm with fluorescing nuclei are observed in the two SST lumina. The lower-middle panel shows a histological section of a portion of a SST containing sperm (the dense rod-like structures in the lumen are sperm nuclei. The arrow indicates the transition between the pseudo-stratified columnar ciliated epithelium of the uterovaginal junction and the simple columnar epithelium of the SST that is characterized in histological preparations by the supra-nuclear vacuole.

The duration of sperm storage in the SSTs is species-dependent. Chickens can store sperm for up to three weeks, whereas turkeys can maintain sperm for 10 weeks in the SST and still lay a fertilized ovum [27-28]. This may be related to number of SSTs present in the UVJ; turkeys have been reported to have 20,000-30,000 SSTs, while chickens have been estimated to have only 5,000-13,500 [29-30]. Additionally, after several generations of selection for high fertility, chicken hens possessed increased numbers of SSTs when compared to non-selected control hens, suggesting the number of SSTs may be positively correlated with fertility [31]. In contrast, under commercial conditions, different broiler strains exhibiting different fertility levels revealed similar numbers of SSTs [29].

Little is known concerning the cellular and molecular mechanisms that sustain sperm within the SST lumen for prolonged periods of storage. These mechanisms likely involve the reversible suppression of sperm motility and metabolism, protection and repair of the sperm plasma membrane, uptake and storage of molecules to sustain sperm metabolism, and maintenance of the SST lumen by removing by-products of sperm metabolism and degraded sperm [32-33]. It is clear the SSTs generate a discrete environment to maintain sperm viability via the influx and efflux of compounds critical for sperm survival [25, 34). While ultrastructural analysis has revealed only limited evidence of secretory activity [25], the identification of membrane-bound vesicles released from the apical tips of the SST epithelial cell microvilli suggests a role in the maintenance of resident sperm through lipid transfer [22, 25, 26,32, 35, 36]. A large proportion of the sperm plasma membrane is composed of polyunsaturated fatty acids [37] that are highly susceptible to damage induced by lipid peroxidation [37]. The peroxidation of these fatty acids results in increased damage to and permeability of the sperm plasma membrane [39, 40]. A complex system of anti-oxidation enzymes are present in the SST epithelial cells and presumably interact with luminal sperm to minimize damage due to lipid peroxidation and maintain sperm membrane integrity [41]. While many metabolites required by sperm in the SSTs have yet to be identified, increased avidin expression is apparent in SSTs relative to surrounding UVJ epithelial tissue possibly providing a means of sequestering biotin and other vitamins for use by the SSTs or resident sperm [42-43]. Interestingly, progesterone has been shown to induce expression of avidin in the oviduct, providing a potential link between progesterone fluctuation and sperm storage in and release from the SSTs [42, 44, 45].

Sperm exit the SSTs in a slow, continuous stream [46-49]; however, a stimulus cuing the egress of resident sperm from the SSTs has yet to be identified. The observations that receptors for estrogen and progesterone exist in the SSTs has led to the suggestion that these compounds may trigger release of resident sperm, possibly in response to hormonal cues over the course of the ovulatory cycle[50-52]. However, an alternate theory suggests the inherent mobility of the sperm plays a larger role than hormonal induction in egress of sperm from SSTs [9]. Resident sperm exhibit a slow, synchronized oscillatory movement in the lumen of SSTs, suggesting the presence of a fluid current through the SST lumen [23-24]. The identification of water channels, known as aquaporins, in the apical epithelium of SSTs lends credence to a model wherein motile sperm maintain their residence in the SST lumen by swimming against the fluid current generated via the aquaporins [53-56]. In the SST lumen, sperm retain their motility by fatty acid oxidation. It has been suggested the sperm membrane is the source of this fatty acid and that as the quality of the sperm membrane gradually decreases there is a reduction of available ATP and sperm motility decreases [56]. Sperm are then swept out of the SST lumen into the UVJ, where they encounter various stimuli enhancing their motility. These sperm are then transported to the infundibulum, the site of fertilization [57]. Such motility-enhancing factors may include changes in environmental pH and neuroendocrine factors such as serotonin [58-62]. Further oxidation of sperm fatty acids, possibly sequestered from the surround milieu, generates the energy required for sperm to respond to such motility-enhancing factors and transcend the oviduct [9,22, 55, 63].

Once sperm are deposited in the oviduct, several selection barriers must be overcome prior to ascending to the infundibulum and fertilizing an ovum. This selection occurs initially in the vagina: only highly mobile (defined as progressive movement in a viscous medium at 40°C) sperm traverse the vagina [9]. While sperm mobility is a major factor in sperm selection in the vagina, sperm selection is also dependent upon the glycoprotein composition of the sperm plasma membrane. The sperm glycocalyx is highly complex and heavily sialylated and modification of the glycocalyx results in reduced fertility and failure of the sperm to enter the SSTs [64-67]. Interestingly, removal of membrane-associated carbohydrates did not affect sperm entry into SSTs if sperm were inseminated directly into the UVJ or when co-incubated with UVJ explants, suggesting the glycocalyx plays a central role in sperm transport and selection through the vagina [64, 66, 68]. Further barriers to sperm prior participating in the process of fertilization include sperm release from the SST and subsequent transport to the infundibulum, and their interaction with the ovum (reviewed in [69]).

## 2.4. Sperm: Ovum Interaction And Fertilization

Given the voluminous nature of the hen's ovum and the GD relative to mammalian ova, one must assume that yet to be identified factors "attract" sperm to the GD. Examination of the electrophoretic profile of the GD and non-GD regions of the PL revealed no variation in protein composition [70]. Furthermore, the abrogation of the preferential interaction of sperm and the inner PL overlying the GD *in vitro* suggests the factors underlying the preferential binding of sperm are not necessarily associated with the inner PL [70]. It is clear, however, glycoproteins play a large role in the interaction between the sperm and ova, even if not directly involved in targeting of sperm to the GD *in vivo* [71]. Pre-treatment of either the PL or sperm with N-glycanases resulted in significantly decreased sperm-ovum interaction *in vitro* [68, 71]. Conversely, N-linked oligosaccharides released from the inner PL by N-glycosidase treatment could induce the acrosome reaction in sperm *in vitro* [72]. These findings strongly suggest N-linked glycans, most likely terminal N-acetyl glucosamine residues, have an essential role in the sperm-ovum interaction in avian species, specifically in induction of the acrosome reaction [68, 72].

Interaction between the sperm and inner PL results in induction of the acrosome reaction [73]. During the acrosome reaction, the inner and outer acrosomal membranes dehisce resulting in the release of acrosin (a trypsin-like enzyme) [21, 74]. As the result of the acrosome reaction, sperm hydrolyze a small hole in the inner PL (Figure 5), enabling sperm to reach the microvilli-studded surface of the ovum [21, 74]. The capacity of sperm to hydrolyze and penetrate the inner PL is the biological basis for the sperm penetration assay discussed below and next section.

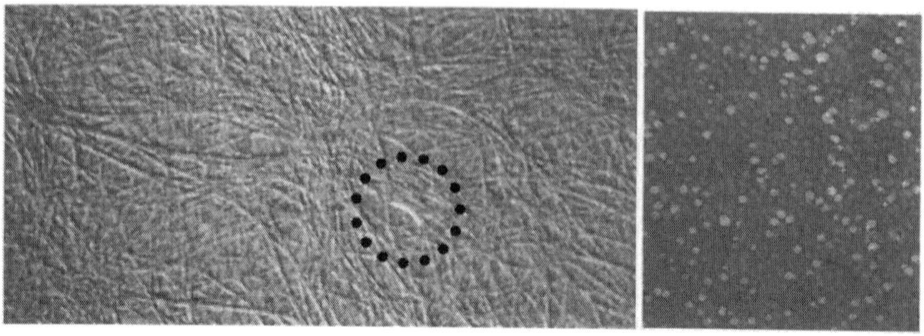

**Figure 5.** In the left panel, a turkey sperm stained with Hoechst 33342 prior to insemination is observed on the surface of the inner perivitelline layer (PL). The sperm's acrosome will release a trypsin like enzyme, acrosin, and digest a hole through the inner PL. The right panel shows multiple sperm holes (white perforations) in the inner PL overlying the germinal disc (GD) of a duck ovum (polyspermy is normal in birds). Sperm hole numbers can be used to assess true fertility and the duration of the fertile period..

Unlike mammals, polyspermy is the norm in avian fertilization. The GD (3.5 mm in diameter) provides a relatively small target for fertilization in the large megalecithal ova (yolk-filled ova) of chickens and turkeys (3.5 – 4.0 mm in diameter); thus polyspermy may be an evolutionary adaptation to ensure higher rates of fertilization in such species [74]. The inner PL may be penetrated by many sperm, although only one male pronucleus will ultimately fuse (syngamy) with the female pronucleus to form the nascent embryo (reviewed in [75-77]. A single sperm hole in the inner PL does not ensure fertilization. Although turkeys show a lower number of sperm interacting with ova relative to chickens, the presence of three sperm holes in the inner PL predicts a 50% probability of fertilization, whereas, six sperm holes suggest a probability greater than 95% fertilization [78]. The outer PL is rapidly depositied around the ovum in the posterior infundibulum and proximal magnum and is impenetrable by sperm [21, 78-79] thus preventing pathological polyspermy.

Given the volume of the GD relative to a single sperm, another possible function of polyspermy may be to activate specific molecular factors in the GD cytoplasm thereby initiating the process of embryogenesis. Yet, polyspermy also results in the presence of multiple male pronuclei in the GD. To cope with this potentially harmful scenario, the mature ovum has been found to have DNase I and II endonuclease activities, both of which will degrade sperm DNA [76]. In contrast, no such DNase activity has been detected in mammalian ova that engage in monospermic fertilization, further suggesting the role of these enzymes in the avian embryo is to protect against detrimental genetic consequences of polyspermy [76].

The number of holes in the inner PL is highly positively correlated with fertility. Correlations exist between the number of sperm inseminated, the number undergoing the acrosome reaction at the inner PL [80], and the number of sperm embedded in the outer PL [81]. The number of sperm holes in the inner PL and the number sperm trapped in the outer PL may be used to estimate the duration of fertility ('fertile period') in hens. While the number of sperm penetrating the inner PL shows a decreasing logarithmic relationship over time [81-82], a positive correlation between the total number of sperm penetrating the inner PL and the number of sperm stored in the SSTs was observed [83]. Given these observations, it should not be surprising there is also a positive correlation between the number of SSTs containing sperm and the proportion of sperm that have undergone the acrosome reaction at the inner PL [82].

# 3. TECHNIQUES IN ARTIFICIAL INSEMINATION AND FERTILITY EVALUATION IN POULTRY

For non-domestic birds, chapters in Bakst and Long [4], Lake and Stewart [5] and Bakst and Wishart [6] provide overviews of semen evaluation and AI techniques. Artificial insemination technology and reproductive biology for ratites were reviewed by Malecki et al. [84].

## 3.1. Semen Collection

Primarily due to the anatomical variation of the phallic region in different birds, semen collection techniques will vary. In contrast to ratites and water-fowl with an intromittent phallus, Galliformes (chicken, turkey, and quail) do not have an intermittent organ. Their non-intromittent organ consists of folds and bulges that make contact with the female's cloaca at mating. From an anatomical perspective, there are considerable differences between the non-intromittent organs of the chicken and turkey (Figure 6). The rooster has a prominent medial phallic body and relatively small lateral phallic bodies and lymph folds. Conversely, the turkey tom has no medial phallic body but prominent lateral phallic bodies and lymph folds. Sex sorting at hatch by cloacal examination is based on the relative differences in size of these structures between the males and females.

The goal of semen collector is to obtain the maximum volume of clean, high quality semen with the minimal amount of handling. In chickens and turkeys, the abdominal massage technique [1, 4] involves massaging the cloacal region to achieve phallic tumescence. This is followed by a 'cloacal stroke', a squeezing of the region surrounding the sides of the cloaca to express the semen. Little additional semen can be expressed after two cloacal strokes; additional cloacal strokes may cause damage to the phallic and cloacal regions and contribute to semen contamination [85].

Semen should be pearly white, viscous, and clean. With each male collected, the semen collector should perform a visual examination of the semen at the time of ejaculation. This is easier with the turkey because the ejaculate accumulates on the phallus before it is collected by the 'milker' (semen collector). Off-color or watery semen, and semen contaminated with blood or fecal/urates debris should not be used for insemination. Due to the increased volume of transparent fluid in rooster semen, which is a transudate derived from the phallus at the time of ejaculation, chicken semen is less viscous and sperm concentration lower than that of turkey semen.

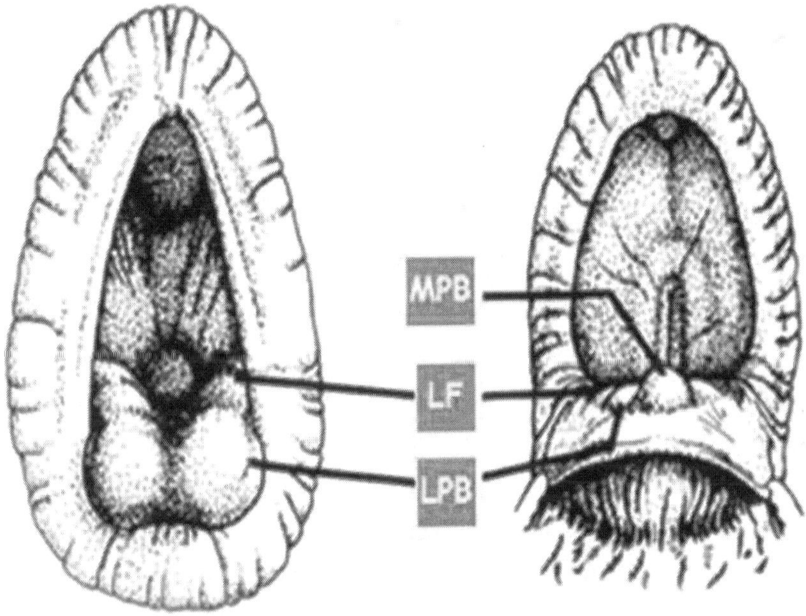


**Figure 6.** The turkey (left) and chicken (right) cloacae are viewed with the dorsal lips of the cloacae pulled back to expose each species phallus non-protrudens. Unlike the turkey, the chicken has a central protuberance, the medial phallic body (MPB) and regressed lateral phallic bodies (LPB) and lymph folds (LF). The turkey phallus non-protrudens is characterized by dominant LPB and LF and the conspicuous absence of the MPB.

## 3.2. Sperm Concentration

If semen is to be diluted, it is best .to have a known volume of semen diluent (a tissue culture-like medium formulated to sustain sperm viability) at ambient temperature in the semen receptacle before collection begins. For routine AI of turkey hens, semen from 10-12 toms are pooled in a single receptacle, mixing the semen gently after each male is collected. Semen volume is determined and if the AI dose is based on numbers of sperm (generally 250-350 million sperm per dose) sperm concentration is determined. The most popular techniques for determining sperm concentration are the packed cell volume (PCV; also referred to as a spermatocrit) and optical density (OD; photometry).

Determining sperm concentration using PCVs is nearly identical to that of determining blood hematocrit values. Semen aspirated into micro-hematocrit tubes are centrifuged in a hematocrit centrifuge until the sperm are tightly packed (10 min); the percentage of packed sperm cells relative to the original semen volume in the micro-tube is determined. Sperm concentration is derived using a conversion factor or standard curve previously derived by comparing and graphically plotting varying ascending sperm concentrations from hemocytometer counts to corresponding spermatocrit readings. (See [4] for detailed protocols to determine sperm concentration and the derivation of standard curves.)

The optical density (OD) is determined using a photometer. The OD of highly diluted semen is directly proportional to the concentration of sperm, thus providing an indirect estimate of the sperm concentration. Like the PCV method, sperm concentration is derived using a conversion factor or previously derived standard curve by comparing and graphically plotting varying sperm concentrations from hemocytometer counts to corresponding OD readings [4].

The PVC and OD methods are two *indirect* methods of determining sperm concentration, that is, the final concentration is calculated from a regression equation or standard curve derived, in part, from*direct* sperm counts with a hemocytometer [4]. Briefly, to derive a regression equation and standard curve, serial dilutions (n=5) covering a wide range of sperm concentrations are prepared and sperm concentrations are determined with a hemocytometer and the instrument or method that requires the standard curve (at least 4 replicates with 4 different semen samples). This is a tedious procedure but if reliable and repeatable sperm numbers are to be inseminated it is best to establish standard curves for each instrument every 12-18 months. The reason for this is that the rotational speed of different centrifuges and the intensity of a photometer's light source may differ as a result of manufacturer's variation, age of the instrument, and/or repeated use of the instrument, thereby producing variations in the respective final readings and subsequent calculations of sperm concentrations.

Another concern when using any semen evaluation method is variation in the operator's techniques. Consistency is the key to repeatable data. The technical staff all must follow the same standard operating procedures (SOPs). For example, when counting sperm with a hemocytometer, all individuals in a lab should following the same SOP for how long the sperm are permitted to settle on the grid and which sperm to count or omit from the count. Also, is the photometer zeroed with the same buffer? If a procedure calls for an incubation period, such as in a live-dead stain, are the samples being incubated for the same duration each time using the same stain concentrations? A lack of consistency in following the SOPs within a laboratory will lead to unwarranted variation and non-reproducible and inaccurate data.

## 3.3. Sperm Viability

In the context of semen evaluation, reference to 'viable' sperm simply implies that such sperm possess an intact plasmalemma and are assumed to be functional. Plasmalemma integrity is frequently determined using either a dead-cell or a live-cell stain alone or simultaneously. The dead-cell stains are excluded by sperm with an intact plasmalemma but stain dead sperm possessing a permeable plasmalemma. Live-cell stains permeate the intact sperm plasmalemma and become visible only after reacting with cytosolic enzymes or interacting with sperm nuclear proteins. Both eosin and propidium iodide are popular dead-cell stains while calcein AM and SYBR-14 are frequently used live-cell stains (see [86] for extended discussion and availability for the live-cell probes). On a commercial breeder farm, the nigrosin/eosin (N/E) technique is most likely the procedure to be used to determine sperm viability [4]. Briefly, sperm are stained with N/E and a smear of the stained sperm is made on a slide (Figure 7). Under a bright field microscope the viable sperm remain pearly white, while eosin will stain non-viable sperm a pink to magenta color. The nigrosin serves as a background to enhance differentiation between the non-viable and viable sperm. In contrast to the N/E technique, a more sophisticated laboratory may use flow cytometry that sorts viable from non-viable sperm after staining with calcein AM or SYBR-14 and propidium iodide.

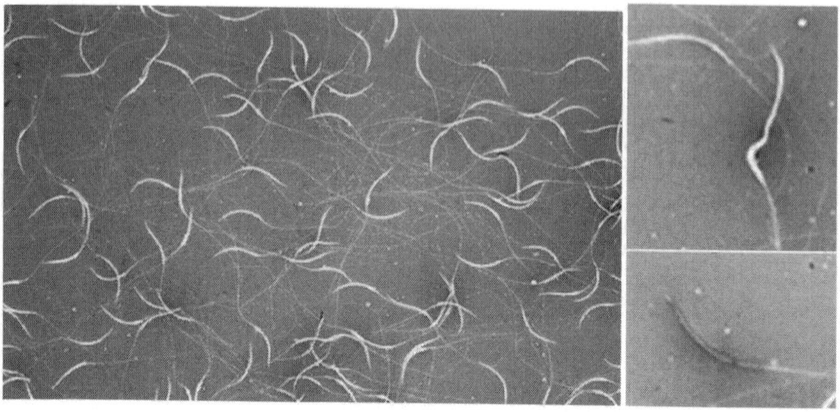

**Figure 7.** The left panel shows a nigrosin eosin preparation of turkey sperm with nearly 100% viable sperm (unstained) white nuclei and midpieces. The sperm head is clearly visible as the white arcing segment; the acrosome and midpiece are difficult to differentiate from the nucleus. The upper right panel reveals a normal sperm and a second sperm with an abnormally curved and swollen midpiece. Observed in the lower right panel is a nonviable sperm stained with eosin throughout the nucleus and midpiece. Barely visible at the anterior end of the nucleus is the unstained, conical shaped acrosome.

## 3.4. Sperm Motility And Mobility

Sperm motility can be progressive (forward direction) or non-progressive (random movement or oscillations) movement. Generally, progressive motility is determined subjectively at ambient temperature using a microscope at low magnification (hanging-drop technique) or objectively using a computer-assisted semen analysis system. These techniques are reviewed by Bakst and Long [4]. Motility evaluated by microscopy has been shown to have little correlation with fertility and simply reveals that the sperm are motile. First described by Froman and McLean [87] and further elaborated for commercial use by Froman [88], the sperm mobility assay has gained popularity as a measure of an individual male's ability to produce highly mobile sperm [mobility defines the ability of sperm to move progressively against a viscous medium (Accudenz) at 41°C] that are more likely to fertilize an ovum than males producing less mobile sperm. While the sperm mobility assay is a powerful tool for the selection of the most fecund males to be used in AI, it necessitates attention to details and accurate and consistent preparation of the reagents.

## 3.5. Evaluation Of Fertility

The measure of a successful AI program is sustained hen fertility. While candling-fertility is useful, there is an eight or more day lag between the last AI and candling-fertility determination, which overlaps with the next insemination (hen insemination is generally at 7-day intervals). With AI programs, it is often desirable to determine the fertility status of a flock before the next weekly insemination. There are several options available: breaking-out fresh eggs and examining the GD to differentiate a fertilized from an unfertilized or early dead embryo; setting normal but culled eggs (checked, hairline cracked, or dirty eggs) in a spare incubator for 24-36 hr before breaking-out [89]; counting sperm in the outer PL; and counting sperm holes in the inner PL. The above procedures are reviewed in Bakst and Long [4].

As noted previously, the sperm penetration assay is not only used to determine true fertility, but also to estimate the number of sperm residing in the use SSTs at the time of ovulation [90]. The isolation of the inner PL and staining procedure, initially developed for chicken eggs by Bramwell et al. [91], was quickly adapted to turkey eggs by Donoghue [92]. The major drawback to the sperm penetration assay as originally described is that it is time consuming, particularly with respect to isolating, washing, and positioning the PL wrinkle-free on the slide. Spasojevic [10] and colleagues at Willmar Poultry Company (Willmar, MN) significantly increased the efficiency of preparing the PL slides from turkey eggs in the following manner: the albumen is removed from the ovum as in the original procedure [4]; a square is outlined on a slide using super glue; the slide is placed firmly on the ovum's surface with the GD centered in the square; after the glue is set, the PL is cut and washed to remove adhering yolk. The advantages here are speed and the PL remains wrinkle free.

A different modification of the sperm penetration assay was suggested by I.A. Malecki (personal communication) and entails placing a filter ring over the GD (inside diameter slightly larger than the GD), cutting around the outside diameter of the filter ring (about 2 mm between the inside and outside perimeter of the ring), and lifting the filter ring off the ovum. The filter ring with the adhering PL is washed gently with saline to remove the yolk and GD material until transparent, placed on a slide, and then fixed and stained with saline washes after each step. Our laboratory has used the filter ring technique with eggs from broilers, turkeys, ducks, and quail and it is now our preferred method for the performing the sperm penetration assay.

# 4. CONCLUSION

Artificial insemination is a common practice in the poultry industry with the turkey industry in North America and Europe using it almost exclusively for the production of hatching eggs. The broiler industry has not adapted AI for several reasons: because of sheer numbers of broiler breeders that need to be inseminated weekly, the labor cost would be very significantly; the initial investment in special housing for the males; an efficient, cost effective means of actually performing the inseminations (housing and catching the hens) would need to be developed; and finally, the concern that after a few generations of breeding broilers by AI, the behaviors associated with natural mating may be less dominant. Notwithstanding these concerns, the benefits of AI for broilers would include the following: the male:female ratio would be increase from 1:10 for natural mating to 1:25 with AI; with fewer males needed, there would be greater selection pressure on the male traits of economic importance and subsequently greater genetic advancement per generation; biosecurity concerns associated with "spiking" aging hen flocks with new and/or younger males to augment mating frequency and fertility would be eliminated; and, differences in body conformation between males and females that impact semen transfer at mating would no longer be a consideration.

In 1995, Sir Peter Lake wrote an excellent review of the history of AI, its impact on the poultry industry, and what is needed to advance the practice of AI with poultry [93]. Unfortunately, AI technology has not advanced significant since this review article. More than 15 years later, the only significant advance is in the evaluation of sperm mobility and the impact that males producing highly mobile sperm have on paternity [9]. Notwithstanding, it is foreseeable that sometime in the future, research addressing poultry sperm biology and the cellular and molecular basis of oviductal sperm transport, selection, and storage will lead to the following innovations in AI technology: insemination intervals increased to 10-14 days (versus 7-day) with fewer sperm per insemination; *in vitro* sperm storage for 24-36 hr at ambient temperature with minimal loss of sperm viability; and, the possibility of transgenic progeny following the insemination of sperm carrying transgenes.

## REFERENCES

1 Burrows WA, Quinn JP. The method of obtaining spermatoza from the domestic fowl. Poultry Science 1935;14(4) 251-4.

2 Quinn JP, Burrows WA. Artificial insemination in fowls. The Journal of Heredity 1936;27(1) 31-7.

3 Spasojevic R. Two hundred million sperm cells per hen? No Way! *Proceedings – Midwest Poultry Federation Convention*, Turkey Breeder Workshop, March 16-18, 2010. CD available at MPF. St. Paul, Minnesota. www.midwestpoultry.com

4 Bakst MR, Long JA. Techniques for Semen Evaluation, Semen Storage, and Fertility Determination. 2, 1-113. 2010. St. Paul, Minnesota, The Midwest Poultry Federation.

5 Lake PE, Stewart JM. Artificial insemination in poultry. Bulletin 213. 1978. London, UK, Ministry of Agriculture, Fish, and Food. Her Majesty's Stationary Office.

6 Bakst MR, Wishart GJ. Proceedings, First Internation Symposium on the Artificial Insemination of Poultry. 1, 1-297. 1995. Champaign, IL, Poultry Science Association, Inc.

7 Bakst MR, Cecil H. Techniques for Semen Evaluation, Semen Storage, and Fertility Determination. 1-97. 1997. Savoy, IL, Poultry Science Association.

8 Jamieson BGM. Reproductive Biology and Phylogeny of Birds. Enfield, New Hampshire: Science Publishers; 2007.

9 Froman DP, Feltmann AJ, Pendarvis K, Cooksey AM, Burgess SC, Rhoads DD. A proteome-based model for sperm mobility phenotype. Journal of Animal Science 2011;89(5) 1330-7.

10 Bakst MR. Physiology and Endocrinology Symposium: Role of the oviduct in maintaining sustained fertility in hens. Journal of Animal Science 2011;89(5)1323-9.

11 Perry MM, Waddington D, Gilbert AB, Hardie MA. Growth rates of the small yolky follicles in the ovary of the domestic fowl. IRCS Medical Science: Experimental Animals 1983;11(11/12) 979-80.

12 Saito N, Sato K, Shimada K. Prostaglandin levels in peripheral and follicular plasma, the isolated theca and granulosa layers of pre- and postovulatory follicles, and the myometrium and mucosa of the shell gland (uterus) during a midsequence-oviposition of the hen (Gallus domesticus). Biology of Reproduction 1987;36(1) 89-96.

13 Dick HR, Culbert J, Wells JW, Gilbert AB, Davidson MF. Steroid hormones in the postovulatory follicle of the domestic fowl (*Gallus domesticus*). Journal of Reproduction and Fertility 1978;53(1) 103-8.

14 Hammond RW, Olson DM, Frenkel RB, Biellier HV, Hertenlendy F. Prostaglandins and steroid hormones in plasma and ovarian follicles during the ovulation cycle of the domestic hen (*Gallus domesticus*). General and Comparative Endocrinology 1980;42(2) 195-202.

15 Hertenlendy F, Biellier HV, Todd H. Effects of the egg cycle and the route of administration on prostaglandin-induced oviposition of hens and Japanese quail. Journal of Reproduction and Fertility 1975;44(3) 579-82.

16 Johnson AL, Woods DC. Ovarian dynamics and follicle development. In: Jamieson BGM, editor. Reproductive Biology and Phylogeny of Birds.Enfield, New Hampshire: Science Publishers; 2007. p. 243-77.

17 Wyburn GM, Johnston HS, Draper MH, Davidson MF. The fine structure of the infundibulum and magnum of the oviduct of *Gallus domesticus*. Quarterly Journal of Experimental Physiology 1970;55(3):213-32.

18 Perry MM. Nuclear events from fertilisation to the early cleavage stages in the domestic fowl (*Gallus domesticus*). Journal of Anatomy 1987;150(1) 99-109.

19 Birkhead TR, Sheldon BC, Fletcher F. A comparative study of sperm–egg interactions in birds. Journal of Reproduction and Fertility 1994;101(2) 353-61.

20 Bakst MR, Howarth B, Jr. Hydrolysis of the hen's perivitelline layer by cock sperm in vitro. Biology of Reproduction 1977;17(3) 370-9.

21 Okamura F, Nishiyama H. The passage of spermatozoa through the vitelline membrane in the domestic fowl, Gallus gallus. Cell and Tissue Research 1978;188(3) 497-508.

22 Bakst MR, Wishart G, Brillard JP. Oviductal sperm selection, transport, and storage in poultry. Poultry Science Reviews 1994;5 117-43.

23 Froman DP, Feltmann AJ, McLean DJ. Increased fecundity resulting from semen donor selection based upon in vitro sperm motility. Poultry Science 1997;77(1) 73-7.

24 Bakst MR. Observations on the turkey oviductal sperm-storage tubule using differential interference contrast microscopy. Journal of Reproduction and Fertility 1992;95(3) 877-83.

25 Schuppin GT, Van Krey HP, Denbow DM, Bakst MR, Meyer GB. Ultrastructural analyses of uterovaginal sperm storage glands in fertile and infertile turkey breeder hens. Poultry Science 1984;63(9) 1872-82.

26 Bakst MR, Akuffo V. Alkaline phosphatase reactivity in the vagina and uterovaginal junction sperm-storage tubules of turkeys in egg production: Implication for sperm storage. British Poultry Science 2007;48(4) 515-8.

27 Brillard JP. Sperm storage and transport following natural mating and artificial insemination. Poultry Science 1993;72(5) 923-8.

28 Christensen VL, Bagley LG. Efficacy of fertilization in artificially inseminated turkey hens. Poultry Science 1989;68(5) 724-9.

29 Bakst MR, Donoghue AM, Yoho DE, Moyle JR, Whipple SM, Camp MJ, et al. Comparisons of sperm storage tubule distribution and number in 4 strains of mature broiler breeders and in turkey hens before and after the onset of photostimulation. Poultry Science 2010;89(5) 986-92.

30 Birkhead TR, Moller AP. Numbers and size of sperm storage tubules and the duration of sperm storage in birds: A comparative study. Biological Journal of the Linnean Society 1992;45(4):363-72.

31 Brillard JP, Beaumont C, Scheller MF. Physiological responses of hens divergently selected on the number of chicks obtained from a single insemination. Journal of Reproduction and Fertility 1998;114(1) 111-7.

32 Bakst MR. Oviducal sperm selection, transport, and storage in poultry: A review. Reproduction Fertility and Development 1993;5(6) 595-9.

33 Blesbois E, Brillard JP. Specific features of *in vivo* and *in vitro* sperm storage in birds. Animal 2007;1(10):1472-81.

34 Van Krey HP, Ogasawara FX, Pangborn J. Light and electron microscopic studies of possible sperm gland emptying mechanisms. Poultry Science 1967;46(1) 69-78.

35 Fujii S. Histological and histochemical studies on the oviduct of the domestic fowl with special reference to the region of utero-vaginal junction. Archives of Histology and Cytology 1963;23(5) 447-9.

36 Renden JA, May EB, Benoff FH. Histochemistry of uterovaginal sperm-host glands in Japanese quail (*Coturnix coturnix japonica*) with reference to the period of oviposition. Poultry Sci 1981;60(11) 2529-35.

37 Blesbois E, Lessire M, Grasseau J, Hallouis JM, Hermier D. Effect of dietary fat on the fatty acid composition and fertilizing ability of fowl semen. Biology of Reproduction 1997;56(5) 1216-20.

38 Surai PF. Natural Antioxidants in Avian Nutrition and Reproduction. Nottingham, UK: Nottingham University Press; 2002.

39 Williams K, Frayne J, Hall L. Expression of extracellular glutathione peroxidase type 5 (GPX5) in the rat male reproductive tract. Molecular Human Reproduction 1998;4(9) 841-8.

40 Catalá A. Lipid peroxidation modifies the picture of membranes from the "Fluid Mosaic Model" to the "lipid whisker model". Biochimie 2012;94(1) 101-9.

41 Breque C, Surai P, Brillard JP. Roles of antioxidants on prolonged storage of avian spermatoza *in vivo* and *in vitro*. Molecular Reproduction and Development 2003;66(3) 314-23.

42 Long C, Sonstegard TS, Long JA, van Tassell CP, Zuelke KA. Serial analysis of gene expression in turkey sperm storage tubules in the presence and absence of resident sperm. Biology of Reproduction 2003;69(2) 469-74.

43 Foye-Jackson OT, Long JA, Bakst MR, Blomberg LA, Akuffo V, Silva VB, et al. Oviductal expression of avidin, avidin-related protein-2, and progesterone receptor in turkey hens in relation to sperm storage: Effects of oviduct tissue type, sperm presence, and turkey line. Poultry Science 2011;90(7) 1539-47.

44 O'Malley BW, McGuire WL. Studies on the mechanism of action of progesterone in regulation of the synthesis of specific protein. Journal of Clinical Investigation 1968;47(3) 654-64.

45 Korpela JK, Elo HA, Tuohimaa PJ. Aviadin induction by estrogen and progesterone in te immature oviduct of chicken, Japanese quail,

duck, and gull. General and Comparative Endocrinology 1981;44(2) 230-2.

46 Burke WH, Ogasawara FX. Presence of spermatozoa in uterovaginal fluids of the hen at various stages of the ovulatory cycle. Poultry Science 1969;48(2) 408-13.

47 Compton MM, Van Krey HP, Siegel PB. The filling and emptying of the uterovaginal sperm-host glands in the domestic hen. Poultry Science 1978;57(6) 1696-700.

48 Bushman AF, Van Krey HP, Denbow DM, Siegel PB. Effect of the ovulatory cycle on oviductal sperm storage in the domestic fowl. Theirogenology 1985;23(3) 473-9.

49 Bakst MR. Duration of fertility of turkeys inseminated at different times after the onset of photostimulation. Journal of Reproduction and Fertility 1988;84(2) 531-7.

50 Das SC, Nagasaka N, Yoshimura Y. Changes in the expression of estrogen receptor mRNA in the utero-vaginal junction containing sperm storage tubules in laying hens after repeated artifical insemination. Theriogenology 2006;64(4) 893-900.

51 Yoshimura Y, Koike K, Okamoto T Immunolocalization of progesterone and estrogen receptors in the sperm storage tubules of laying and diethylstilbestrol-injected immature hens. Poultry Science 2000;79(1) 94-8.

52 Ito T, Yoshizaki N, Tokumoto T, Ono H, Yoshimura Y, Tsukada A, et al. Progesterone is a sperm-releasing factor from the sperm-storage tubules in birds. Endocrinology 2011;152(10) 3952-62.

53 Agre P, Sasaki S, Chrispeels MJ. Aquaporins: A family of water channel proteins. American Journal of Physiology Renal Physiology 1993;265(3) F461.

54 Zaniboni L, Bakst MR. Localization of aquaporins in the sperm storage tubules in the turkey oviduct. Poultry Science 2004;83(7) 1209-12.

55 Froman DP, Bowling ER, Wilson JL. Sperm mobility phenotype not determined by sperm quality index. Poultry Science 2003;82(3) 496-502.

56 Froman DP. Effect of sperm mobility phenotype on fertility, sperm competition, and in vivo sperm storage in the domestic fowl. Journal of Dairy Science 93[Suppl. 1], 504. 2010.

57 Brillard JP, Galut O, Nys Y. Possible causes of subfertility in hens following insemination near the time of oviposition. British Poultry Science 1987;28(2) 307-18.

58 Stephens RE, Prior G. Dynein from serotonin-activated cilla and flagella: Extraction characteristics and distinct sites for cAMP-dependent protein phosphorylation. Journal of Cell Science 1992;103(4):999-1012.

59 Holm L, Ridderstrale Y, Knutsson PG. Localisation of carbonic anhydrase in the sperm storing regions of the domestic hen. Cell Tissues Organs 1996;156(4):253-60.

60 Holm L, Wishart GJ. The effect of pH on the motility of spermatoza from chicken, turkey and quail. Animal Reproduction Science 1998;54(1):45-54.

61 Freedman SL, Akuffo VG, Bakst MR. Evidence for the innervations of sperm storage tubules in the oviduct of the turkey (*Meleagris gallopavo*). Reproduction 2001;121(5):809-14.

62 Bakst MR, Akuffo V. Serotonin localization in the turkey vaginal but not sperm storage tubule epithelia. Journal of Avian Biology 2008;39(3):348-54.

63 Froman DP, Bowling ER, Wilson JL. Sperm mobility phenotype not determined by sperm quality index. Poultry Science 2003;82(3):496-502.

64 Froman DP, Engel HN. Alteration of the spermatozoal glycocalyx and its effect on duration of fertility in the fowl (*Gallus domesticus*). Biology of Reproduction 1989;40(3):615-22.

65 Froman DP, Thursam KA. Desialylation of the rooster sperm's glycocalyx decreases sperm sequestration following intravaginal insemination of the hen. Biology of Reproduction 1994;50(5):1094-9.

66 Steele MG, Wishart GJ. Demonstration that the removal of sialic acid from the surface of chicken spermatozoa impedes their transvaginal migration. Theriogenology 1996;46(6):1037-44.

67 Pelaez J, Long JA. Characterizing the glycocalyx of poultry spermatozoa: I. Identification and distribution of carbohydrate residues using flow cytometry and epifluorescence microscopy. Journal of Andrology 2007;28(2):342-52.

68 Robertson L, Wishart GJ, Horrocks AJ. Identification of perivitelline N-linked glycans as mediators of sperm-egg interaction in chickens. Journal of Reproduction and Fertility 2000;120(2):397-403.

69 Birkhead TR, Brillard JP. Reproductive isolation in birds: postcopulatory prezygotic barriers. Trends in Ecology & Evolution 2007;22(5):266-72.

70 Steele MG, Meldrum W, Brillard JP, Wishart GJ. The interaction of avian spermatozoa with the perivitelline layer in vitro and in vivo. Journal of Reproduction and Fertility 1994;101(3):599-603.

71 Howarth B. Carbohydrate involvement in sperm-egg interaction in the chicken. Journal of Receptor Research 1992;12(2):255-65.

72 Horrocks AJ, Stewart JM, Jackson L, Wishart GJ. Induction of acrosomal exocytosis in chicken spermatozoa by inner perivitelline-derived N-linked glycans. Biochemical and Biophysical Research Communications 2000;278(1):84-9.

73 Koyanagi F, Masuda S, Nishiyama H. Acrosome reaction to cock spermatoza incubated with the perivitelline layer of the hen's ovum. Poultry Science 1988;67(12):1770-4.

74 Bakst MR, Howarth B, Jr. The fine structure of the hen's ovum at ovulation. Biology of Reproduction 1977;17(3):361-9.

75 Wasserman PM, Jovine L, Litscher ES. A profile of fertilization in mammals. Nature Cell Biology 2001;3(2):E59-64.

76 Stepinska U, Olszanska B. DNase I and II present in avian oocytes: A possible involvement in sperm degradation at polyspermic fertilisation. Zygote 2003;11(1):35-42.

77 Stepinska U, Bakst MR. Fertilization. In: Jamieson BGM, editor. Reproductive Biology and Phylogeny of Birds. Enfield, New Hampshire: Science Publishers; 2007. p. 553-87.

78 Wishart GJ. Quantitative aspects of sperm:egg interaction in chickens and turkeys. Animal Reproduction Science 1997;48(1):81-92.

79 Bellaris R, Harkness M, Harkness RD. The vitelline membrane of the hen's egg: A chemical and electron microscopal study. Journal of Ultrastructure Research 1963;8(3):339-59.

80 Bramwell RK, Howarth B, Jr. Preferential attachment of cock spermatozoa to the perivitelline layer directly over the germinal disc of the hen's ovum. Biology of Reproduction 1992;47(5):1113-7.

81 Wishart GJ. Regulation of the length of the fertile period in the domestic fowl by numbers of oviducal spermatozoa, as reflected by those trapped in laid eggs. Journal of Reproduction and Fertility 1987;81(2):495-9.

82 Brillard JP, Antoine H. Storage of sperm in the uterovaginal junction and its incidence on the numbers of spermatoza present in the perivitelline layer of hen's eggs. British Poultry Science 1990;31(3):635-44.

83 Wishart GJ. Quantitative aspects of sperm:egg interaction in chickens and turkeys. Animal Reproduction Science 1997;48(1):81-92.

84 Malecki IA, Rybnik PK, Martin GB, Malecki I, Glatz P. Artificial insemination technology for ratites: A review. Australian Journal of Experimental Biology 2008;48(10):1284-92.

85 Bakst MR, Cecil HC. Gross appearance of turkey cloacae before and after single or multiple manual semen collections. Poultry Science 1983;62(4):683-9.

86 Haugland RP. Handbook of Fluorescent Probes and Research Products. 9th ed. Eugene, Oregon: Molecular Probes, Inc.; 2002.

87 Froman DP, McLean DJ. Objective measurement of sperm motility based upon sperm penetration of Accudenz®. Poultry Science 1996;75(6):776-84.

88 Froman DP. Application of the sperm mobility assay to primary broiler breeder stock. Journal of Applied Poultry Research 2006;15(2):280-6.

89 Bakst MR, McGary S, Estevez I, Knapp T. Use of nonsettable eggs to evaluate turkey hen fertility. Journal of Applied Poultry Research 2002;11(4):402-5.

90 Brillard JP, Bakst MR. Quantification of spermatozoa in the sperm-storage tubules of turkey hens and the relation to sperm numbers in the perivitelline layer of eggs. Biology of Reproduction 1990;43(2):271-5.

91 Bramwell RK, Marks HL, Howarth B. Quantitative determination of spermatozoa penetration of the perivitelline layer of the hen's ovum as assessed on oviposited eggs. Poultry Science 1995;74(11):1875-83.

92 Donoghue AM. The effect of twenty-four hour *in vitro* storage on sperm hydrolysis through the perivitelline layer of ovipositioned turkey eggs. Poultry Science 1996;75(8):1035-8.

93 Lake, PE. Historical perspective of artificial insemination technology. In: Bakst MR and Wishart GJ, editors. Proceedings - First International Symposium on the Artifical Insemination of Poultry. Savoy, Illinois: Poultry Science Association; 1995. p. 1-20

# CHAPTER 2

# Farm Models and Eco-Health of Poultry Production Clusters (PPCs) following Avian Influenza Epidemics in Thailand

**Worapol Aengwanich**

Stress and Oxidative Stress in Animal Research Unit, Faculty of Veterinary Sciences, Mahasarakham University, Maha Sarakham 44000, Thailand; Tel./Fax: +66-043-742-823

## ABSTRACT

Thailand is located in Southeast Asia and is a country that was affected by highly pathogenic avian influenza (HPAI) epidemics during 2003–2004. Nevertheless, the Thai government's issuance policy of strict control and prevention of the disease has resulted in efficient disease control of avian influenza (AI). Poultry farmers have been both positively and negatively affected by this policy. There are three poultry cluster models worthy of attention in Thailand: (1) egg chicken poultry clusters over ponds; (2) egg chicken poultry clusters in coops raised from the ground and managed by a cooperative; and (3) poultry clusters in closed coops under contract with the private sector. Following the AI epidemics, additional poultry husbandry and biosecurity systems were developed, thereby generating income and improving the quality of life for poultry farmers. Nevertheless, raising large clusters of poultry in the same area results in disadvantages, particularly problems with both air and water pollution, depending upon the environments of each poultry model. Furthermore, the government's policy for controlling AI during epidemics has had a negative effect on the relationship between officials and farmers, due to poultry destruction measures.

## KEYWORDS

poultry production clusters; PPCs; eco-health; avian influenza (AI); Thailand

## 1. INTRODUCTION

Thailand's population is predominantly agricultural, and most people hold agricultural occupations involving both the cultivation of crops and animal husbandry for domestic consumers. During the period from the 1950s to the 1970s, the Thai government began to have a policy to produce food for export, and poultry products were one of the agricultural product types included in the production plans for export [1]. The aforementioned policy has had a tremendous impact on the nation's poultry farming, as poultry farmers have transitioned from raising poultry in their back yards to industrial poultry farming. Later on, the country's poultry industry experienced huge success [2]. Despite Thailand's high poultry production volume, 90% of the nation's poultry farmers continue to be small-scale farmers [3]. Most of Thailand's poultry farming is done in the country's central and eastern regions [4].

In 2003–2004, Thailand experienced avian influenza (AI) epidemics [5], and most of these occurred on farms with low biosecurity levels. These epidemics had a tremendous impact on poultry production and the living, health and socio-economic conditions of poultry farmers [4]. During the AI epidemics, the Thai government put in place disease control measures, such as stamping out, cleaning and disinfection, surveillance, movement control, campaigns for awareness promotion and panic reduction, improvement of biosecurity and restructuring of poultry production. These resulted in a continuous decline of AI outbreaks. The policy was implemented through four processes, namely quarantine, screening, zoning and compensation [4,6,7,8,9,10,11,12].

Burgos *et al.* [13] studied and classified poultry farming in Thailand into three systems. The first system was the traditional production system. The second system was the semi-industrial one. This type of system was quite large and connected with the commercial production system. The third system was the industrial system. Aengwanich *et al.* [14] defined domestic poultry production clusters (PPCs) as "areas of concentrated poultry production in rural areas usually separated from residential areas—where farms practice certain economies of scale and use comparative advantages of localities, resulting in improvements to and increased intensification of poultry production. The farms in the cluster apply standard biological safety and environmentally friendly practices and install related facilities. Major drives for the formation of PPCs and changes to PPCs include government programs for enhanced bio-security at both national and local levels, extended families, cooperatives and improved links to contracting companies that supply inputs and purchase outputs and outbreaks of highly pathogenic avian influenza".

The concept of eco-health, which focuses on the surrounding ecology, such as the physical environment, economic and social aspects and culture relating to health and the welfare of human beings [15], has played a key role in studies on the impact of the environment on human health. Nevertheless, following the AI epidemics in Thailand, socio-economic data on farming, including the various impacts of AI epidemics on poultry farming models, socio-economic characteristics, disease control and farming environments on small poultry cluster farms, became limited.

Therefore, the objective of the present study was to investigate the socio-economic farming characteristics, disease control and farming environments on small farms in PPCs following outbreaks of AI. The pool of knowledge gained from the present study will be of benefit in setting policies for the prevention and control of AI, improving the quality of life of poultry farmers and reducing problems with pollution in the environments surrounding poultry farms in the future.

# 2. RESEARCH SITES AND METHODOLOGY SECTION

## 2.1. Research Sites

The present study was conducted in three provinces in Thailand, namely, poultry clusters in Nong Khai, Nakhon Phanom and Maha Sarakham. Detailed descriptions of each area follow.

### 2.1.1. Nong Khai Province

Nong Khai province is located in the upper northeastern region. The area is narrow, stretching along the bank of the Mekong River with an approximate length of 210.6 km. On the opposite bank of the Mekong River is Laos. Nong Khai has an area of 3027.280 $km^2$. Its population is 509,870, with a population density of 168.43 people/$km^2$. Nong Khai is 615 km away from Bangkok and 25 km from Vientiane, the capital city of Laos. In the Nong Khai province, the important poultry raised are layer chickens, and most of the farms are in three main districts, *i.e.*, Muang district, Tha Bo district and Sri Chiang Mai district, with approximately 300 farms. The farming pattern is clusters of 30–100 farms that are located away from the communities, dispersed near natural water sources, e.g., large lakes or areas with irrigation systems and drainage systems, or on one's own land. A notable characteristic of the layer chicken production in Nong Khai province is that they keep fish with hens, building coops on fish ponds. The floor of the barn is built to leave holes to allow the feces of the chickens to fall into the ponds, serving as fish feed. The fish they keep are Nile tilapia and fish in the carp family. Thus, the farmers have two sources of income: from the layer chickens and from the fish.

### 2.1.2. Nakhon Phanom Province

Nakhon Phanom province is located in the upper northeastern region. Along the entire east frontier is the Mekong River. Nakhon Phanom is adjacent to Kham Muan province of Laos. Its area is 5512.668 km$^2$. Its population is 704,768, with a population density of 127.85 people/km$^2$. Nakhon Phanom is 740 km away from Bangkok. The poultry producer clusters in Nakhon Phanom are located in the Muang district. There is one cluster of 59 farm members. The distinct characteristic of the clusters is that they raise layer chickens in raised barns. The farmers' farms are on their own land. The cluster is managed as a cooperative group. The members jointly invest to provide factors of production from the outside. The farmers have been raising layer chickens in raised barns for over 20 years. The farmers mix the chicken feed with the material received from the cooperatives. The residential area, feed stock and preparation areas and the egg sorting barns are separated from the chicken barns.

### 2.1.3. Maha Sarakham Province

Maha Sarakham province is located in the center of the northeastern region. The province has an area of 5291.683 km$^2$. Its population is 939,736, with a population density of 177.59 people/km$^2$. Maha Sarakham is 407 km away from Bangkok. The poultry production cluster in Maha Sarakham is in Kosumphisai district, which is adjacent to Khon Kaen province. The distinct characteristic of the farms is that they are situated along the road to the community, with no farms in the community. The farms are closed systems. There are approximately 50 farms in total, most of which keep broilers and female and male layer chickens. The farmers have contracts with private companies and manage their farm as per distinctive agreements. The farmers raise chicken in an evaporative cooling system. Every farm in this system has to be approved for farm standards by the Department of Livestock Development (DLD). The biosecurity system levels are high.

## 2.2. Methodology

This study was a qualitative survey performed during January–August, 2012, after a highly pathogenic avian influenza (HPAI) outbreak in Thailand (the last outbreak of HPAI in Thailand was reported in 2006). The project was approved by the institutional review board (IRB). Data collection was conducted in the three groups of PPC, *i.e.*, farmers who raise poultry over fish ponds in Nong Khai (farmers in this province are small producers and the biosecurity level of poultry farms is low), farmers who raise poultry in the form of cooperatives in Nakhon Phanom (the biosecurity level of poultry farms in this province is medium) and farmers who raise poultry under contract with private companies in Maha Sarakham (all farms were approved for the biosecurity level (high level) by the DLD in Thailand; the details of the biosecurity evaluation are shown in Table 1), in terms of attitudes and behaviors and economic, social, biodiversity, human health, environmental and biosecurity aspects. Data were collected from poultry farmers in the three provinces, amounting to a total of 50 families per province by a structural questionnaire. After that, the researcher conducted interviews with people involved in the PPC in each province, *i.e.*, poultry farmer group leaders, community leaders, sub-district organization administration officials and district and provincial livestock office officials. The researcher also collected data through participative observation of the poultry farmers to conduct an in-depth analysis of the case studies. Next, the data collected were used in focus group discussions to verify the data among the people involved and to conclude the study.

**Table 1.** The biosecurity level of poultry production clusters (PPCs) in Nong Khai, Nakhon Phanom and Maha Sarakham province, Thailand.

| Items | Indicators | NK | NP | MS |
|---|---|---|---|---|
| 1. Components of farm | 1.1. Location | | | |
| | 1.1.1. Farm is located in an environment with a minimal risk from any contamination of physical, chemical and biological hazards. | – | – | + |
| | 1.2. Layout | | | |
| | 1.2.1. Farm shall have a sufficient size, suitable for rearing, and does not pose any problem to the environment. | – | – | + |
| | 1.2.2. The layout of the farm shall be orderly, separated according to farm activities, such as rearing, feed storage, equipment storage, carcass disposal, worker accommodation, egg storage room. | + | + | + |
| | 1.3. Housing | | | |
| | 1.3.1. House shall be strong, hygienic, easily cleaned and maintained and well ventilated. | – | + | + |
| | 1.3.2. The area in the layer house is adequate for hygienic chicken rearing. | – | + | + |
| 2. Feed | 2.1. Feed quality shall comply with the feed quality control act, relevant laws and regulations. | + | – | + |
| | 2.2. All layers can access feed. | + | + | + |
| | 2.3. Feed shall be stored in an area to prevent contamination and deterioration. | + | + | + |
| 3. Watering | 3.1. The water source shall be located in an area safe from the contamination of hazardous substances. | – | + | + |
| | 3.2. The water used shall be cleaned. | – | + | + |
| | 3.3. All layers can access water. | + | + | + |
| 4. Farm management | 4.1. Farm management manual | | | |
| | 4.1.1. The availability of a manual demonstrating the details of the important farm operations, e.g., rearing system, recording system, animal welfare and health management. | + | + | + |

| | | | | |
|---|---|---|---|---|
| | 4.2. Personnel | | | |
| | 4.2.1. The operator who takes care of layer rearing shall be skillful and trained. | ST | ST | + |
| | 4.2.2. The availability of licensed veterinarian to supervise the layer farm and animal health. | – | – | + |
| | 4.3. Cleaning and maintenance | | | |
| | 4.3.1. House and equipment shall be cleaned and maintained in a good and hygienic manner. | ST | ST | + |
| | 4.3.2. Cleaning and disinfection of the house and equipment shall be done after all layers have been removed; and its effectiveness can be examined by swabbing the chicken rearing area for microbial culture. The house shall be vacated according to the notification of the Department of Livestock Development (DLD). | + | + | + |
| | 4.3.3. House litter and nest box litter materials shall be clean and dry. | ST | ST | + |
| | 4.4. Egg management | | | |
| | 4.4.1. The availability of hygienic cleaning and handling of eggs to be properly collected and stored. | – | ST | + |

# 3. RESULTS AND DISCUSSION

Data on social attitudes and behaviors and economic, biodiversity, human health, environmental, and biosecurity aspects obtained from the three groups of poultry farmers, *i.e.*, the group raising poultry over fish ponds in Nong Khai, the cooperative poultry group of Nakhon Phanom and the PPC under contract with private companies in Maha Sarakham, are described below.

## 3.1. Economic Aspects

### 3.1.1. PPC in Nong Khai

The PPC in Nong Khai raised 500–2400 chickens/farm. The case studies revealed that the income from poultry constituted 80.65% of the total household income. Most workers in the production were family members, except for some families with many coops, who had to hire migrant workers from a neighboring country (Lao PDR (People's Democratic Republic)). The farms with a small number of chickens did not hire any employees unless it was necessary. On average, members from the PPC in Nong Khai province took out a loan from the Bank for Agriculture and Agricultural Co-operatives at 360,000 baht/family and had a debt of 320,000 baht/family. Groups of layer chicken farmers in Nong Khai province sold eggs to local middlemen. The eggs produced from the cluster were sold to the middlemen in Nong Khai or nearby provinces, such as Udon Thani, Nong Bua Lamphu and Loei, and the retired hens were sold in the markets and the local slaughterhouse. All transactions were made in cash. Feed, chemical supplies and chicks were provided by local stores. The stores granted credit to the farmers, and there were officers from the DLD training the farmers about disease control.

### 3.1.2. PPC in Nakhon Phanom

The PPC in Nakhon Phanom raised an average of 5000 chickens/household. There were three sources of income for the PPC of Nakhon Phanom province, from selling eggs, chicken manure and retired chickens. The main workers were family members (husband and wife) and other members, e.g., children and in-laws. There were no permanently hired workers. Sometimes, workers were hired for loading chicken manure for sale or some specific tasks, e.g., chicken catching or vaccination. Farmers were paid for the disinfectant and boots, which cost approximately 2000 baht/year. Farmers in this PPC borrowed from the Bank for Agriculture and Agricultural Co-operatives. Debts of 50,000–90,000 baht/household were mostly from feed and medical suppliers for raising chickens. Farmers sold eggs to communities around the cluster and middlemen from the district or nearby provinces. The PPC partnered with a private company that sold food, drugs and medical supplies to the cooperative. The cooperative's members ordered materials from the cluster's center to serve as ingredients for feed or to prevent and control diseases in their farms.

### 3.1.3. PPC in Maha Sarakham

The chickens raised in the PPC in Maha Sarakham were broilers and male and female layer chickens. Each farm had one barn that could accommodate 21,000–28,000 chickens. Female layer chickens could be reared twice per year. Male layer chickens could be reared 3–4 times per year. Poultry farmers made contracts with Sri Viroj and Betagro Company in the form of a contract farming system. Income for the poultry production of farmers in Maha Sarakham province was derived from male and female layer chickens; the average income of each farm was 330,000 baht/year. The average income of farmers came from selling chickens back to the companies, rice farming and sugar cane farming. The workers at the farms were mainly family members. The main labor was from the husband, wife and their children. The number of workers in each family was 2–4 people. There were no hired workers, except for some tasks, e.g., when farmers needed to remove chicken manure, members from the community were hired occasionally, and when it was time to sell chickens, workers from other farms (relatives' farms) were asked to help. Villagers were hired to scoop chicken manure for sale.

At the beginning, farmers built the bathroom, changing room and food storage facilities, bought equipment for disinfectant spraying and boots and fenced the farm. The initial cost was approximately 50,000 baht. The housing was built with a bank loan; afterward, biosecurity measures were put in place. Private companies invested in chicks, feed, drugs, medical supplies, chemicals, veterinary services and other factors of production. The farmers knew only the cost of housing and workers. The other costs were handled by the private companies and were unknown to the farmers. When the chickens gained enough weight, the private companies would come to catch them for weighing and pay the money due to the farmers. Therefore, the farmers did not know the actual cost of the chickens, as they received only the money from farming.

Most farmers borrowed money from the Bank for Agriculture and Agricultural Co-operatives. Each family took out a loan of approximately 300,000–400,000 baht to invest in housing and biosecurity measures. At the time of the study, each farmer had a debt of 100,000–400,000 baht/farm, depending on the loan and payment period.

The PPC in Maha Sarakham did not have any issues with marketing, as they were contract farmers with private companies. The companies bought chickens at the promised rate upon which they had agreed. When it was time, the farmers would sell chickens back to the company.

The positive and negative economic impacts of PPCs on poultry farmers were as follows:

(a) Positive economic impact of PPCs on farmers.

As the measure on poultry production clustering was announced, the obvious positive impact was the control of disease outbreaks within farms, which resulted in an improvement in poultry health. Mortality during poultry rearing decreased, due to more stringent prevention and eradication of disease within the farms. Farmers thus obtained higher poultry productivity. However, this depended on the price of the product at any one time, except in the case of contract poultry farming, in which the income of farmers per unit of output remained constant, and farmers earned more when the productivity was increased or poultry mortality was reduced.

(b) Negative economic impacts of PPCs on farmers.

Higher production costs on housing, the biosecurity system and development toward the adoption of a standard farming system were additional fundamental costs the farmers had to absorb, whereby a loan needed to be obtained from a bank through the use of their own land as a loan guarantee, except in the case where they already had adequate operating capital.

## 3.2. Social Aspects

### 3.2.1. PPC in Nong Khai

For the poultry farming society in Nong Khai at the family level, the labor division among family members was as follows: the father fed and gave water to the chickens, tended the chickens, vaccinated the chickens and gave them medicines, collected eggs, mowed the grass, took care of the fish farming, sold eggs and attended meetings arranged by the DLD; the mother collected and sold eggs, counted the money and did the accounting, placed orders and made payments, helped with the vaccination of chickens, cooked, took care of the children and planted trees; employees (on some farms) helped the father.

The PPC in Nong Khai had continuously undergone economic and social changes, both before and after the AI outbreak, e.g., the economic status of the farmers of Khok Pla Faang village used to be good, and the farmers had better income from chicken farming than from tobacco cultivation. Every family was happy and helpful to others in the village. There were informal talks among the group members, and they could consult with one another as kinsmen, especially about the costs of the factors of production, disease epidemics and chicken husbandry. After the AI outbreak, the economics worsened, and many farms ceased to operate.

The situation was similar with farmers from the layer chicken production cluster in Wieng Khook village. The economics of the cluster were good with many chicken farms in the cluster before the outbreak. After the AI outbreak, the number of farms declined as the price of eggs lowered, whereas the cost soared. The cluster faced problems, and some farms had to close: Khok Pla Faang village was left with only nine farms. Abandoning the chicken farms, some of the farmers continued with tobacco cultivation or other careers, and only fish ponds were left. However, the economics of layer chicken production were considerably better at the time of the study.

Generally, the relationships between farms in Nong Khai province were within the cluster. Most of these were relationships among relatives and a loose relationship between farmers who bought the factors of production from the same store. The latter kind of relationship was loose, with interactions, such as checking the prices of products or a talk at a training session/seminar held by stores or at a meeting arranged by the Provincial Livestock Office. There was not much cooperation among the producers. They checked the prices of feed or how to tend their livestock. In contrast, at the poultry farms in Tha Samraan village, as most farmers were relatives, they usually consulted and helped one another, including providing occasional financial support. Finally, the relationships within the cluster became colder, due to the more demanding work and internal competitiveness. It was obvious that the AI outbreak had severe economic effects on the layer chicken production in this province. Even after the recovery and the measures taken, the economics of farmers had improved, but the relationships within the cluster showed a tendency to worsen.

### 3.2.2. PPC in Nakhon Phanom

The PPC of Nakhon Phanom divided work within the families as follows: the farm owner (male) mixed feed, fed, gave water to and vaccinated the chickens and collected the eggs; his wife sorted and collected the eggs, tended the coops and gave water to the chickens. When more workers were needed, e.g., for vaccination, the entire family would be at work.

The layer chicken production cluster at Nakhon Phanom was a group of farmers in a large community that had been raising layer chickens for almost 20 years. During the first phase, the cluster was bound to the local capitalists for credits for feed, drugs and medical supplies for raising chicken. As they had been in debt from continual farming after the AI outbreak in the year 2006 (some farmers stopped farming then), the farmers grouped to buy feed, chicken breeds and medical supplies, each investing 20,000–50,000 baht. There were occasional group meetings. As their farms were in close proximity, the farmers always consulted with one another.

The PPC in Nakhon Phanom received financial support in forming the cluster from the province via the Livestock Office of Nakhon Phanom province. The farms had no relationships outside the group, except with private companies that provided feed, drugs and medical supplies for internal use.

The farming of the production cluster of Nakhon Phanom province was cooperative. A board of directors was founded and there was investment in the form of stocks for providing factors of production and joint prevention and control of diseases within the community, e.g., disinfectant spraying in the streets and the area of the community. There were meetings among cooperative members to set measures and to plan operations and solve problems within the cluster.

### 3.2.3. PPC in Maha Sarakham

Regarding domestic social relationships, the labor division in the PPC in Maha Sarakham was as follows: the husband brought the feed into the housing; together, the wife and children fed the chickens, spread the manure in the housing, gave water to the chickens, removed the manure and cleaned the housing. The wife also took care of housework.

For inter-farm relationships within the cluster, the chicken farming of the PPC in Maha Sarakham was initiated by the family “Donlakhorn”. This was because the cost of building chicken housing and of running a closed system was high. Later, they supported their relatives and extended family farms. Meanwhile, worthy activities (such as donations to the temple or charity) in the community went on as before. Consultation in the cluster was about environmental issues or problems from raising chicken, e.g., fly elimination or smell removal among clusters in the same networking company once or twice a year. Most of the cooperation of the farmers in the cluster was to solve issues, such as the ammoniac smell and flies in the community when there were complaints. The issues were solved by drying the manure, selling some to other communities as fish food and using fly repellent.

However, they did not associate with farms in other networking companies, even if they were relatives.

## 3.3. Attitudes and Behaviors

### 3.3.1. PPC in Nong Khai

From the past to the present, all poultry houses have been built above fish ponds, because the aforementioned format enables farmers to have multiple sources of income from the sale of eggs, retired hens and fish. Thus, the farmers maintain the aforementioned format, because they deem it suitable for their families. The system of chicken housing on fish ponds in Nong Khai province reduces the ammoniac smell and flies, yet the problems found were that some farms were run with an inappropriate rate of fish, layer chickens and size of fish ponds. Thus, the water became foul, as in the case of Tha Samraan village, where the fish ponds had not been let dry, the fish died and the water became foul from an excessive amount of chicken manure. In terms of general consumption among poultry production farmers in Nong Khai, there was poultry and egg consumption, as per local culture.

In contrast, disease control in the PPC in Nong Khai province was not strict. Only some farms followed a strict and correct method for disease control and treatment. Most of them focused on these only during the epidemic. The required practice was proper vaccination, dipping egg packs and feet (for some farms) in disinfectant, the spraying of disinfectant, disinfecting the barn, prohibiting outsiders from entering the farms (this might have been controlled more strictly during the outbreak), treating chickens with antibiotics and vitamins, removing infected chicken for treatment when found and consulting with other clusters when an outbreak occurred (mostly, treatment would be given only after the disease was found). There was no warning system. Sometimes, the farmers learned from the drug seller about outbreaks in other districts, as each farm bought drugs from the store in the network.

### 3.3.2. PPC in Nakhon Phanom

In general, farmers in the PPC in Nakhon Phanom consumed eggs and cooked retired layer chickens, according the consumption culture in the community. The farmers raised chickens in raised poultry houses, due to limited land and the fact that the people in this province refused to consume fish fed chicken manure. Although certain farms also raised fish, fish sales were faced with a problem of acceptance by people. In terms of disease control, the disease control system for every farm in the cluster was as follows: separating feed and the egg sorting barn; keeping visit records; applying a proper disease control system; using disinfectant pools; changing clothes and shoes before entering the coops; the control of vehicles entering the farm; fencing the farm area; and weekly disinfection of the farm when the weather changed.

### 3.3.3. PPC in Maha Sarakham

The customary culture of the PPC in Maha Sarakham is the consumption of indigenous chickens and ducks, with various poultry cooking methods, including roasting, boiling, stewing and making a spicy minced meat salad. Sometimes, they eat the poultry produced in the farming system, but not so often.

In the community that was separate from the cluster, fighting cocks, indigenous chickens and ducks were kept. In the PPC, the male and female layer chickens were reared in a closed system. Even though the farmers in Maha Sarakham province raised chickens in a modern closed system, they still held onto some traditional chicken raising practices, e.g., giving boiled *Andrographis* or *Tinospora crispa* to chickens to cure the flu.

Farms in Maha Sarakham had a good biosecurity system, *i.e.*, at the gate at the farm entrance, before entering the barn, the farmers would shower, change their clothes and spray; the residential zone and the housing were separate. There was a storage room for materials, and chickens were reared in a closed system. When any chicken became sick, the farmer would inform the company, and the company would send their staff or veterinarians to treat the chicken, giving it drugs, vitamins and vaccines. Flies and smells were controlled by spraying chemicals onto the bedding. The farmers removed chicken manure from the barn and sold it to buyers.

## 3.4. Biodiversity Aspects

According to the data on the biodiversity of poultry raised in Nong Khai, Nakhon Phanom and Maha Sarakham, the PPC in Nong Khai mainly raised layer chickens, but some farms also kept indigenous breeds, ducks and turkeys. Among the exotic breeds found in the PCC were teals, white egrets, other exotic breeds and some local birds. In Nakhon Phanom, the PPC only kept layer chickens. In addition, there were also a few migratory birds and some local birds, as there were trees all around the farms. Poultry breeds found in the PPC in Maha Sarakham were layers, male layers, broilers, indigenous chickens and ducks. A few migratory birds and some local birds were also found, as there were trees all around the farm.

## 3.5. Human and Animal Health Aspects

As the biosecurity aspect of the disease control measures was in place and strictly implemented by the farmers in the PPCs, e.g., appropriate housing improvement, preventing animals from entering the poultry houses, farm entry and exit control, disinfection, the appropriate practices of workers, the appropriate disposal of the remains of dead and sick animals, medication, *etc.*, it was found that the incidences of poultry sickness or death in the farms with strict biosecurity control were reduced significantly. Animal health was improved, and infections in poultry were reduced. Meanwhile, with regard to the health of farmers, the government had a policy for farmers in PPCs, such as those in Nong Khai and Maha Sarakham, to receive a medical examination and influenza vaccination on an annual basis. This was considered a regular health practice for the monitoring and care of farmers.

## 3.6. Environmental Aspects

As PPCs took shape, this meant a collection of a large number of poultry farms in one area, from which environmental impacts arose. The size of the impact depended on the management and control of the contamination of the waste that spread from the farms into the environment, e.g., poultry droppings, ammonia odors developing from accumulated droppings, dust diffusion and a large number of flies in certain seasons. The management of environmental problems in various areas differed based on farming types, housing styles, land characteristics, waste disposal systems, *etc.*

Examples of the management of environmental and pollution issues among the farms in the PPCs were as follows:

(a) Case study of PPC in Nong Khai

The type of farming of layer chicken production clusters in Nong Khai had the unique characteristic that layer houses were built on top of a fish pond, whereby layer chicken droppings fell into the pond and became fish feed. As a result, there was no problem of ammonic odors due to the accumulation of droppings. There were neither flies nor mosquitoes in the farms. As the fish in the pond grew and reached the required weight, however, the water had to be pumped out and released into natural waterways. The water being pumped out contained much organic matter, which gave rise to an impact on the water sources; the impact depended on the volume of the water and the organic matter being released. This farming type, however, did not cause odor problems due to poultry droppings for the surrounding communities.

(b) Case study of PPC in Nakhon Phanom

The rearing of layer chickens in the Nakhon Phanom PPC was done in raised floor housing, where chicken droppings fell onto the ground underneath. After a period of accumulation, absorbing much moisture or becoming wet from rain, ammonia gas developed in a high volume and affected the environment within the farm, the farmers and the neighbors. During certain periods of time, there were also a large number of flies, which caused much annoyance. Farms that employed good disposal methods for chicken droppings did not face the aforementioned problems; on the other hand, if the management was not appropriate, the farmers were often involved in a dispute with community inhabitants.

(c) Case study of PPC in Maha Sarakham

The PPC in Maha Sarakham reared male and female layer chickens and broilers in closed housing with an evaporative cooling system. Therefore, the problems of dust, flies and odor were considerably small, except where rearing density was high and moisture within the poultry houses was also high, due to inappropriate water management. The accumulation of poultry droppings emitted very bad odors. When odorous air circulated outside, it caused air pollution and problems with neighboring farms. However, as the cluster was distant from the communities, the problem was not severe. Moreover, sub-district administration organizations also enforced legislation controlling the construction of poultry houses where farmers wishing to set up a farm had to obtain the consent of the owner of neighboring pieces of land before the construction could proceed. This lessened the problem of odor in communities.

## 3.7. Biosecurity Aspects

During the outbreaks of AI, the Thai government issued various disease control and prevention policies that called for more stringent biosecurity management in poultry farms at all levels. However, after the outbreaks ended, the biosecurity control of certain PPCs relaxed, due to the high investment and complicated procedures. Most farmers in the clusters tended to apply the strict control measures during outbreaks of poultry diseases in their vicinity, such as the PPC in Nong Khai. On the contrary, the biosecurity control system of the PPC in Nakhon Phanom remained in place and was strictly followed by the farmers, because the government authorities had regularly promoted and given advice on farming practices, assessed compliance with farming standards and applied appropriate control strategies, which resulted in strict compliance with the government measures. The PPC in a contract farming system in Maha Sarakham also had the biosecurity system in place, and it was strictly implemented, as all farms had to achieve the farming standard certification, which was one of the conditions of the contracts made with large private companies. All farmers had to be certified, and this resulted in their high standard of disease prevention.

## 3.8. Relationship between Officials and Poultry Farmers in PPCs

During AI epidemics, the government had a policy for the prevention and control of AI with a number of measures, such as stamping out, cleaning and disinfection, surveillance, movement control, the improvement of biosecurity and the restructuring of poultry production. This policy, particularly the destruction of poultry within a 5-km radius of outbreaks of the disease, resulted in tensions in the relationship between officials and poultry farmers, with the poor relationship on a continual decline. Farmers did not dare to report data, because they were afraid the officials would come and destroy the poultry they raised on their farms if they did. Therefore, the aforementioned reasons are indicators of another impact of the policies for the prevention of AI outbreaks, particularly concerning stamping out measures. In other words, the gap between officials and poultry farmers widened.

After the policy to control the spread of AI was announced in Thailand, the DLD, on behalf of the government, issued a disease control notification and instructed all provincial livestock offices to abide by the measures of which they were notified, whereby all poultry farms had to implement the measures to prevent and control disease. As a result of AI outbreaks, farmers were required to adopt the farming standard that called for standardized disease prevention and control. For those groups of poultry farmers who had already formed a cluster, e.g., the PPC in Nong Khai, the provincial livestock office promoted and encouraged farmers to increase the biosecurity level of each layer chicken farm. Activities included the improvement of poultry housing sanitation, the installation of netting around the housing, the implementation of a system for preventing pathogens from entering the farm, such as showering, wearing boots and changing clothes, the installation of fencing to prevent outsiders from entering the farm and the control of transportation.

In the case of the PPC in Nakhon Phanom, the farmers had changed from traditional rearing to a common form of farming that was a cooperative group and upgraded their farms to standard poultry farms, which had to be certified by the DLD, with a focus on the improvement of housing and the implementation of biosecurity and environmental systems. Farmers had to invest in such improvements by themselves, on top of the original investment made to develop the PPC system in accordance with government policy.

The PPC in Maha Sarakham was the group where changes in the grouping pattern of poultry farmers had been clear-cut based on the government AI outbreak control policy: before the AI outbreaks, farmers used to rear poultry, such as broilers, under contract with large private companies, such as Charoen Pokphand (CP) affiliates, whereby a semi-intensive farming system was implemented. After the AI outbreaks, other large private companies approached with a new farming system by encouraging the clustering of layer chicken farms, in which 30–40 chicken farms were located in the same area. Farmers had to invest in housing by obtaining a bank loan to construct the closed system housing and to apply for farm standard certification from the DLD. The farming system changed from a guaranteed price system to a hired farming system, such as in the case of layer chicken farming to raise and deliver pullets to the company that was a party to the contract, whereby the company would supply the farmers with such production inputs as chicks, feed, chemicals, biological products, e.g., vaccines, as well as pay electricity bills for the closed system operation. When the contract term was due, the company would repay the wage and rental charge for the housing and cages.

As a result of the change to comply with the AI control policy, the direct economic impact on poultry farmers was the investment in a new farming system, which might include the improvement of housing, the rearing system, the biosecurity system, *etc.* In the case where the farmers did not have adequate funds to continue their operation after being affected by multiple measures of the government during AI outbreaks, some had to discontinue farming, due to the lack of funds for ongoing operation and/or for upgrading the farming system to the level required by the government policy.

The Thai social structure comprised the communities and the people therein who were related through the kinship system, as every member of a family needed to share the use of land resources, which had been shared from generation to generation. As the community structure changed, joint decisions within the family or community were usually required, as could be seen from the Nong Khai PPC, where groups of poultry farmers mainly developed in the form of extended family farms. However, the encouragement to restructure the grouping other than as kinship groups usually arose from external factors, such as government support. This form of social relationship, on the other hand, was still based on kinship systems. An example is the setup of the poultry production cooperative in Nakhon Phanom: although this was a social group established through government support, it was based on the kinship system and the relationships among neighbors in the same community.

When analyzing the overall characteristics of PPCs in Thailand, however, it was found that in the setup of the clusters, each farm was located on its own land, except in the case of compartment-type farming, where large private companies bought the land and set up farms of their own. The administrative activities of farmers in PPCs were therefore complete in their own farms and were commercially independent of neighboring farms. This might be different from China or Vietnam, where the government arranged production inputs, such as land, which made it possible to gather a group of farmers to form a production cluster in the same area.

In general, the poultry production methods of Thai farmers involve back yard poultry production with no technology or a planned poultry production model, particularly with small farmers in rural areas. Hence, the poultry production of farmers is only a supplemental occupation or food source. Back yard poultry production began to transform into poultry production for sale or commercial purposes, but knowledge about strict or intensive poultry production relied on new knowledge and technology, including strict disease prevention and control. Consequently, the poultry farmers either did not understand or were unable to practice the aforementioned measures. Therefore, when poor disease control was practiced on the farms, outbreaks of disease occurred at those farms on a regular basis and resulted in disadvantages for the poultry farmers as more chickens became ill and income decreased. In addition, because the farmers had no understanding of the importance of preventing and controlling disease, they had few ideas about developing or improving biosecurity systems. Hence, when there were outbreaks of AI in Thailand in 2003–2004, even though the government had a policy in place for poultry farms to develop biosecurity systems that the poultry farms had to follow (in the early stages, farmers went along with the policy), farmers' adherence to the policy for preventing and controlling the disease eased when there were fewer AI outbreaks. Therefore, measures for instruction, building correct understanding and adjusting attitudes and behaviors are essential in developing disease prevention and control on poultry farms.

Former poultry production models in PPCs in the province of Nong Khai involved raising chickens in pens raised high above the ground. Later on, however, poultry production methods involved experiments with raising chickens over fish ponds, which yielded positive results, with increased income from the fish. Moreover, the people in this province accept and consume fish fed chicken excrement as the main food source. Therefore, poultry production models for egg-laying chickens in Nong Khai are all based on the same design, which differs from Nakhon Phanom, where most people do not consume fish fed chicken excrement; hence, the fish raised by this method are usually unsellable. Therefore, the only poultry production method in the province of Nakhon Phanom involves raising poultry in pens raised above the ground.

Nevertheless, although poultry farmers were found to raise chickens with more modern systems in the case of the PPC in the province of Nakhon Phanom, farmers were still found to use former ideologies in raising chickens, such as using herbs to treat sick chickens, which indicated that communities possessed original knowledge and continued to believe in the use of former practices in raising chickens in combination with poultry production for commercial purposes or based on modern methods, because herbal methods continued to yield good outcomes.

In terms of biosecurity, the findings indicated that raising large numbers of birds on the same farm put the farm at high risk for outbreaks of disease. Poultry farms raising poultry by modern methods raised only one type of poultry on the farm. According to the findings of the present study, the PPC in Nong Khai raised other types of poultry, such as ducks, geese or turkeys, in addition to egg-laying chickens on the same farm, while clusters of farmers raising egg-laying chickens in Nakhon Phanom raised only one breed of egg-laying chickens. The same was true with the PPC in Maha Sarakham. Therefore, the PPC in Nong Khai was at a higher risk for outbreaks of disease than PPCs in Nakhon Phanom and Maha Sarakham.

According to the policy for increasing biosecurity systems on poultry farms, more biosecurity systems have been employed on poultry farms. Wei and Aengwanich [17] conducted a study on biosecurity systems by employing an evaluation form and found that biosecurity systems for PPCs in Maha Sarakham, which had been accredited by the DLD, were of higher quality than those in Nong Khai and Nakhon Phanom. Furthermore, the increase in biosecurity systems on the farms reduced the number of poultry deaths in the PPCs in Maha Sarakham and made the number of poultry deaths lower than that in Nong Khai and Nakhon Phanom. In addition, because AI is a disease of epidemic proportions among animals that spreads from animals to humans, the government also has a policy for controlling and preventing the disease in humans [11]. Therefore, farmers whose work involves exposure to poultry and officials receive health check-ups. Moreover, these poultry farmers are vaccinated against influenza annually in order to help them build immunity against the disease.

Poultry farms generally encounter air environmental issues, due to fermented chicken excrement and the accumulation of ammonia gas in large amounts. The aforementioned gas has the effect of irritating the eyes and respiratory tracts of poultry farmers/workers. Furthermore, exposure to the gas can cause respiratory diseases [18,19,20]. According to the findings of the present study, the farms in the PPC in Maha Sarakham where chickens were raised in closed systems experienced problems with polluted air, particularly concerning accumulated ammonia gas in the chicken pens. In addition, farms in the PPC in Nakhon Phanom were found to have the same problems. Therefore, it is evident that the ecosystems of the PPCs are constantly confronted with air pollution issues. Hence, measures need to be developed to solve the problem of disposing of large amounts of waste on the farms and solving excrement fermentation problems that lead to odors, such as water dripping over the excrement due to broken watering equipment or leaky roofs during the rainy season, *etc.*

Nevertheless, the aforementioned problems were not encountered on the farms of poultry farmers in the PPC in Nong Khai, where chickens were raised over fish ponds. When the chickens excreted waste, it fell into the ponds and became fish food. Hence, the problem of odor was reduced or eliminated altogether. However, in cases where the chickens excreted waste into the pond in large amounts, but the pond size and number of fish were disproportionate, problems occurred with wastewater in the ponds. Additionally, when the fish grew to the size where they could be caught and sold, farmers released the water from the ponds into public water sources, which resulted in problems with polluted water sources in communities and conflicts between farm owners and community residents. Therefore, problem-solving methods should involve adjustments in the ratio of the number of chickens raised to ensure that it is proportionate to the pond size and the number of fish released into the ponds. Otherwise, the amount of oxygen might be increased in the water sources by using proper equipment.

Biosecurity systems are systems that help prevent diseases from invading farms. According to the findings of the present study, the PPC in each cluster model had a different biosecurity system. In the early days following the policy for disease control on farms in PPCs, the PPC farmers in Nong Khai practiced compliance with government policy. Then, when the outbreaks of the disease had died down, the stringency in adhering to disease control activities gradually declined, while the PPC in Nakhon Phanom continued to practice disease control activities and biosecurity systems, because the DLD officials continued to visit, offering supervision and support of the cluster activities. The two aforementioned phenomena are indications that officials play an important role in assisting farmers in PPCs to realize the importance of disease prevention or biosecurity systems on farms, while the poultry farmers in the PPC in Maha Sarakham required accreditation for compliance with farming standards in strict disease control before they could sign contracts with private sector companies. Hence, the biosecurity systems on these farms were more advanced than those on the farms in PPCs in Nong Khai and Nakhon Phanom. Therefore, it is evident that government mechanisms to support farmers in maintaining standards in biosecurity systems continue to hold great importance for PPC farmers.

During the outbreaks of AI, the government policy requiring the destruction of poultry within a 5-km radius damaged the relationship between government officials and farmers compared with that preceding the AI outbreaks, because the farmers were afraid government officials would come and destroy poultry when their poultry on the farm fell ill. Thus, the farmers did not dare to report data to the officials. Consequently, the role of the officers in visiting to work with the farmers diminished. An obvious consequence was that the PPC in Nong Khai had low biosecurity systems in comparison with the PPC in Nakhon Phanom, because government officials were unable to visit and supervise farmers sufficiently, and the relationship between the officials and farmers was poor. Therefore, the solution is to create a process in which both government officials and farmers can take part.

Eco-heath is a concept linking the environment, socio-economics and health [21]. According to the findings of the current study, when the concept of eco-health is combined with the analysis of PPCs in Thailand, the results in the field are in accordance with the eco-health concept in terms of environment, because PPCs in the same area generally encounter problems in both air and water pollution. In addition, the incomes and lifestyles of farmers in these PPCs are involved with the effects on the relationship between government officials and the farmers when there are new outbreaks of AI on the farms or in nearby areas. Consequently, farmers are required to invest in additional disease prevention. Furthermore, when poultry die, the farming economy is affected. In addition, the related economics are also affected at the micro and macro levels. Moreover, outbreaks of diseases occurring in these poultry can infect humans. Therefore, it is evident that studies on all three dimensions will offer an aggregate perspective of the problems and enable thorough prevention of their potential impact, including recommendations for policy guidelines for the future.

## 4. CONCLUSIONS

The following three types of PPCs exist in Thailand: (1) PPCs raising chickens over fish ponds; (2) PPCs composed of egg-laying chickens managed in the form of cooperatives; and (3) PPCs in closed pens under contract with private sector companies. Following outbreaks of AI in each type of PPC, the disease prevention systems have been improved on the farms. The income earned from poultry products has increased, and some of the clusters have joined together in the form of cooperatives. Furthermore, there are poultry production systems under contract with private sector companies. However, having PPCs with a large number of farms in the same area has resulted in consequences, particularly air and water pollution, depending on the environments of each type of poultry production. Moreover, the government's policy for AI disease control by poultry destruction during outbreaks of the disease has damaged the relationship between farmers and government officials, thereby preventing the officials from visiting farms to promote poultry production and disease control methods as well as they should.

## ACKNOWLEDGMENTS

This research was funded by The International Development Research Center, Foreign Affairs, Trade and Development Canada (through the Global Health Research Initiative) and the Australian Agency for International Development.

## CONFLICTS OF INTEREST

The author declares no conflict of interest.

## REFERENCES

1 Christensen, S. The Role of Agribusiness in Thai Agriculture: Toward a Policy Analysis. *TDRI Quart. Rev.* **1992**, *7*, 3–9.

2 Borelli, A.; Ghobadi, N.; Otte, J.; Pfeiffer, D.; Roland-Holst, D. *Policy Responses to the Threat of Highly Contagious Disease Originating in Domestic Animals: An Overview of Recent Experience with Avian Influenza*; the Pro-Poor Livestock Policy Initiative; FAO: Rome, Italy, 2005; p. 4.

3 Heft-Neal, S.; Kahrl, F.; Otte, J.; Roland-Holst, D. *HPAI Risk Reduction and Smallholder Poultry Supply Chains—The Case of Thailand. Controlling Avian Flu and Protecting People's Livelihoods in the Mekong Region*; HPAI Research Brief/No. 14; International Food Policy Research Institute: Washington, DC, USA, 2009; pp. 1–4.

4 Tiensin, T.; Nielen, M.; Songserm, T.; Kalpravidh, W.; Chaitaweesub, P.; Amonsin, A.; Chotiprasatintara, S.; Chaisingh, A.; Damrongwatanapokin, S.; Wongkasemjit, S.; *et al*. Geographic and Temporal Distribution of Highly Pathogenic Avian Influenza a Virus (H5N1) in Thailand, 2004–2005: An Overview. *Avian Dis.* **2007**, *51*, 182–188.

5 Verbiest, J.P.; Castillo, C.N. *Avian Flu: An Economic Assessment for Selected Developing Countries in Asia*; ERD Policy Brief No. 24; Asian Development Bank (ADB): Manila, Philippines, 2004; p. 1.

6 Rushton, J.; Viscarra, R.; Bleich, E.G.; McLeod, A. *Impact of Avian Influenza Outbreaks in the Poultry Sectors of Five South East Asian Countries (Cambodia, Indonesia, Lao PDR, Thailand, Viet Nam) Outbreak Costs, Responses and Potential Long Term Control*; TCP/RAS/3010; FAO: Rome, Italy, 2005; pp. 7–8.

7 Tiensin, T.; Chaitaweesub, P.; Songserm, T.; Chaisingh, A.; Hoonsuwan, W.; Buranathai, C.; Parakamawongsa, T.; Premashthira, S.; Amonsin, A.; Gilbert, M.; *et al.* Highly Pathogenic Avian Influenza H5N1, Thailand, 2004. *Emerg. Infect. Dis.* **2005**, *11*, 1664–1672.

8 Barbazan, P.; Thitithanyanont, A.; Misse, D.; Dubot, A.; Bosc, P.; Luangsri, N.; Gonzalez, J.P.; Kittayapong, P. Detection of H5N1 Avian Influenza Virus from Mosquitoes Collected in an Infected Poultry Farm in Thailand. *Vector-Borne Zoonot.* **2008**, *8*, 105–109.

9 NaRanong, V. Structural Changes in Thailand's Poultry Sector: Avian Influenza and Its Aftermath. *TDRI Quart. Rev.* **2008**, *23*, 3–10.

10 Heft-Neal, S.; Kahrl, F.; Otte, J.; Roland-Holst, D. *Assessment of Smallholder Indigenous Poultry Producer Viability in Thailand*; Mekong Team Working Paper No. 9; Department for International Development: London, UK, 2009; pp. 1–5.

11 Safman, R.M. *The Political Economy of Avian Influenza in Thailand*; STEPS Centre: Brighton, UK, 2009; pp. 5–24.

12 Prakarnkamanant, A.; Mastin, A.; Patanasatienkul, T.; Kasemsuwan, S.; Wongsathapornchai, K.; Chanachai, K.; Otte, J.; Pfeiffer, D. *Quantitative Risk Assessment of HPAI Virus H5N1 Release via Cock Fighting Activities into the 1-km Buffer Zones Surrounding Compartmentalised Broiler Chicken Farms in Thailand*; Mekong Team Working Paper No. 11; Department for International Development: London, UK, 2010; pp. vi–x.

13 Burgos, S.; Otte, J.; Metras, D.P.R.; Kasemsuwan, S.; Chanachai, K.; Heft-Neal, S.; Roland-Holst, D. *Poultry, HPAI and Livelihoods in Thailand—A Review*; Mekong Team Working Paper No. 4; Department for International Development: London, UK, 2008; pp. 24–33.

14 Aengwanich, W.; Intarakhamhaeng, M.; Wandee, J.; Nongbua, T.; Chaiyasak, S.; Srikot, P.; Thammasar, K.; Junsanitsri, N.; Sritongtuam, K.; Tawinwaang, T. Poultry Production Clusters

(PPCs) After AI Outbreaks in Thailand: Past, Present and Future Direction. *Int. J. Poultry Sci.* **2012**, *11*, 541–550.

15 Plaen de, R.; Kilelu, C. From Multiple Voices to a Common Language: Ecosystem Approaches to Human Health as an Emerging Paradigm. *EcoHealth* **2004**, *1*, 8–15.

16 National Bureau of Agricultural Commodity and Food Standards, Ministry of Agriculture and Cooperative. Good Agricultural Practices for chicken breeder farm. Available online: www.acfs.go.th/standard/download/eng/GAP_chicken.pdf (accesse d on 14 December 2014).

17 Wei, H.; Aengwanich, W. Biosecurity Evaluation of Poultry Production Cluster (PPCs) in Thailand. *Int. J. Poultry Sci.***2012**, *11*, 582–588.

18 Wheeler, E.F.; Weiss, R.W.J.; Weidenboerner, E. Evaluation of Instrumentation for Measuring Aerial Ammonia in Poultry Houses. *J. Appl. Poultry Res.* **2000**, *9*, 443–452.

19 Ritz, C.W.; Fairchild, B.D.; Lacy, M.P. Implications of Ammonia Production and Emissions from Commercial Poultry Facilities: A Review. *J. Appl. Poultry Res.* **2004**, *13*, 684–692.

20 Rylander, R.; Carvalheiro, M.F. Airways Inflammation among Workers in Poultry Houses. *Int. Arch. Occup. Environ. Health* **2006**, *79*, 487–490.

21 Charron, D.F. *Ecohealth Research in Practice*; The international Development Research Centre: Ottawa, ON, Canada, 2012; pp. 1–30.

# CHAPTER 3

# Poultry Farm Vulnerability and Risk of Avian Influenza Re-Emergence in Thailand

**Marc Souris [1,2,3,†,*], Dubravka Selenic [1,†], Supaluk Khaklang [1], Suwannapa Ninphanomchai [1], Guy Minet [1], Jean-Paul Gonzalez [1,4] and Pattamaporn Kittayapong [1]**

[1] Center of Excellence for Vectors and Vector Borne Diseases, Faculty of Science, Mahidol University at Salaya, Nakhon Pathom 73170, Thailand

[2] UMR_D 190 "Emergence des Pathologies Virales", IRD French Institute of Research for Development, Aix-Marseille University, EHESP French School of Public Health, Marseille 13385, France

[3] Asian Institute of Technology, P.O. Box 4, Klong Luang, Pathumthani 12120, Thailand

[4] Metabiota, Washington, DC 20009, USA

## ABSTRACT

Highly pathogenic avian influenza (HPAI) remains of concern as a major potential global threat. This article evaluates and discusses the level of vulnerability of medium and small-scale commercial poultry production systems in Thailand related to avian influenza virus re-emergence. We developed a survey on 173 farms in Nakhon Pathom province to identify the global level of vulnerability of farms, and to determine which type of farms appears to be more vulnerable. We used official regulations (the Good Agricultural Practices and Livestock Farm Standards regulations) as a reference to check whether these regulations are respected. The results show that numerous vulnerability factors subsist and could represent, in case of HPAI re-emergence, a significant risk for a large spread of the disease. Bio-security, farm management and agro-commercial practices are particularly significant on that matter: results show that these practices still need a thorough improvement on a majority of farms. Farms producing eggs (especially duck eggs) are more vulnerable than farms producing meat. Those results are consistent with the type of farms that were mostly affected during the 2004–2008 outbreaks in Thailand.

## KEYWORDS

HPAI H5N1; vulnerability assessment; poultry production system; Thailand

## 1. INTRODUCTION

The first outbreaks of highly pathogenic avian influenza caused by viruses of the H5N1 subtype (H5N1 HPAI) were detected in geese in 1996 in the Chinese province of Guangdong and in 1997 in Hong Kong. Until today, 54 countries in Asia, Europe and Africa have reported H5N1 HPAI outbreaks among poultry; outbreaks in poultry farms have been reported in every country in South East Asia (SEA), except Brunei, Singapore and the Philippines. A number of preventive measures have been implemented to keep the disease under control; however, it reappears sporadically and is now enzootic in some countries [1]. H5N1 HPAI remains of concern as a major potential global threat.

Thailand officially declared it first H5N1 HPAI outbreak in a layer farm in the central region on January 2004. From 2004 until 2008,

H5N1 HPAI was laboratory confirmed in 1,890 poultry flocks in Thailand, and 17 human deaths out of 25 cases were reported [2]. More than 63 million birds had been culled due to control measures. The epidemics had been determined as a major threat to the poultry industry and caused reduction in chicken production and consumption with a huge economic impact [3]. The last occurrence of H5N1 HPAI outbreak in Thailand was declared to the World Organization for Animal Health (OIE) in November 2008, and, since then no laboratory-confirmed outbreaks have been reported. Nevertheless, studies in ducks revealed serological evidence of H5 infection in 2010 [4], and in some neighboring countries re-emergence of H5N1 HPAI is still ongoing.

Thailand's poultry industry has grown rapidly over the twenty years leading up to 2004, especially in the broiler production area. In 2002, the country was the fourth largest exporter of poultry meat [5]. The poultry production system in Thailand can be classified according to Food and Agriculture Organization (FAO) categorization as: sector 1, large-scale industrial production, with high bio-security; sector 2, semi-industrial production (commercial poultry production system with moderate to high bio-security and birds/products usually marketed commercially); sector 3, small to medium scale semi-commercial poultry production (with low to minimal bio-security and birds/products usually marketed at local bird markets) and; sector 4, smallholder backyard farming with minimal or no bio-security [6].

Vulnerability factors associated with HPAI outbreaks have been the subject of many investigations. Agro-commercial activities, human behaviors, transports and poultry smuggling have been identified as major factors of local dispersion of the virus in Thailand [2,7,8,9,10,11]. To certificate safety poultry production and to minimize poultry farms vulnerability, the Bureau of Livestock Standard and Certification of Department of Livestock Development (DLD, Ministry of Agriculture, Bangkok, Thailand) set standards for Good Agricultural Practice (GAP) and Livestock Farm Standards (LFS), with regular surveillance and yearly recertification audits of poultry farms. Variables based on GAP and LFS comprise eight main sections (for broiler and meat ducks farms) or nine (for layer farms): general condition of the farm, poultry feed, water quality, farm management, animal health, poultry welfare management, environmental management, data recording and productivity management for layers farms. Immediately after the first wave of HPAI outbreaks, in 2004, DLD was immensely concerned in increasing farms bio-security and preparing new policy measures [12]. For example, farmers were required to improve open poultry housing to a closed evaporative system. However these measures were not fully implemented, and still, in certified poultry farm standards, both (open and closed-evap) houses are allowed [13]. No study evaluating the actual level of vulnerability of poultry farms in Thailand using official GAP and LFS regulations was published recently. A study was performed in 2011 in three provinces, but focused on the main pathway of entry of influenza viruses onto farms with different production methods [14].

This article describes and discusses the actual level of vulnerability of medium and small-scale poultry commercial production system (sector 2 and 3) related to HPAI risk in Thailand. We use official GAP and LFS regulations as the references to identify the overall level of vulnerability of poultry farms, checking on farms whether these regulations are respected. The aim of this article is to evaluate the global level of vulnerability, to highlight the most common factors of vulnerability, and to determine which type of farms appears to be more vulnerable. Also, spatial analysis is performed in order to find any geographical pattern of vulnerability susceptible to increase the risk of HPAI farm-to-farm transmission.

# 2. MATERIAL AND METHODS

## 2.1. Field Survey in Nakhon Pathom Province

We conducted a field survey supported by a comprehensive semi-structured questionnaire in the 173 registered commercial poultry farms in the Nakhon Pathom province, Thailand (Figure 1) from March to September 2011. The Nakhon Pathom province, in the central region of Thailand, has been selected for our study because of its extensive agriculture activities and for logistic reasons. The province (2,168 km$^2$) consists of seven districts, 106 sub-districts and 870 villages, with nearly 1,000,000 inhabitants in 2011. Nakhon Pathom is nationally ranked 7th and 2nd for its broiler and duck meat production, 6th and 8th for chicken and duck layer farm (DLD, 2008). It has 212 registered poultry farms belonging to sector 1 to 3 (DLD, 2010). Farms belonging to sector 1 were not included in the study as they are considered to have limited vulnerability because they implement high level operating procedures to reduce the risk of viral incursions [6]. The sector 4 farms (backyard production with minimal bio-security and birds/products consumed locally) were also not included in this study as we considered that outdoor backyard production with minimal or no bio-security measures cannot be equally compared to farms [9,15].

The present survey included all 173 registered commercial poultry farms of the province belonging to sector 2 and sector 3 in 2011: 96 broiler farms, 33 farms of laying hens, 28 ducks meat farms and, 16 farms of laying ducks. The location of each farm entrance was recorded by GPS. An environmental description was conducted from the farm entrance. A questionnaire was used to describe agricultural practices from owners or workers; all interviews were conducted in Thai language. Questionnaircs were addressed directly, without an appointment, to owners or workers, at the entrance of the farms.

## 2.2. HPAI in the Nakhon Pathom Province, 2004–2008

Forty one H5N1 HPAI laboratory confirmed outbreaks were reported in Nakhon Pathom from 2004 to 2005 (Figure 2). The province has suffered strong economic effects, like the neighboring province of Suphanburi where H5N1 HPAI emerged in late 2003 [2]. Of the 173 farms from sector 2 and 3 of Nakhon Pathom inspected in 2011, 43 were established after 2004 (Figure 3).

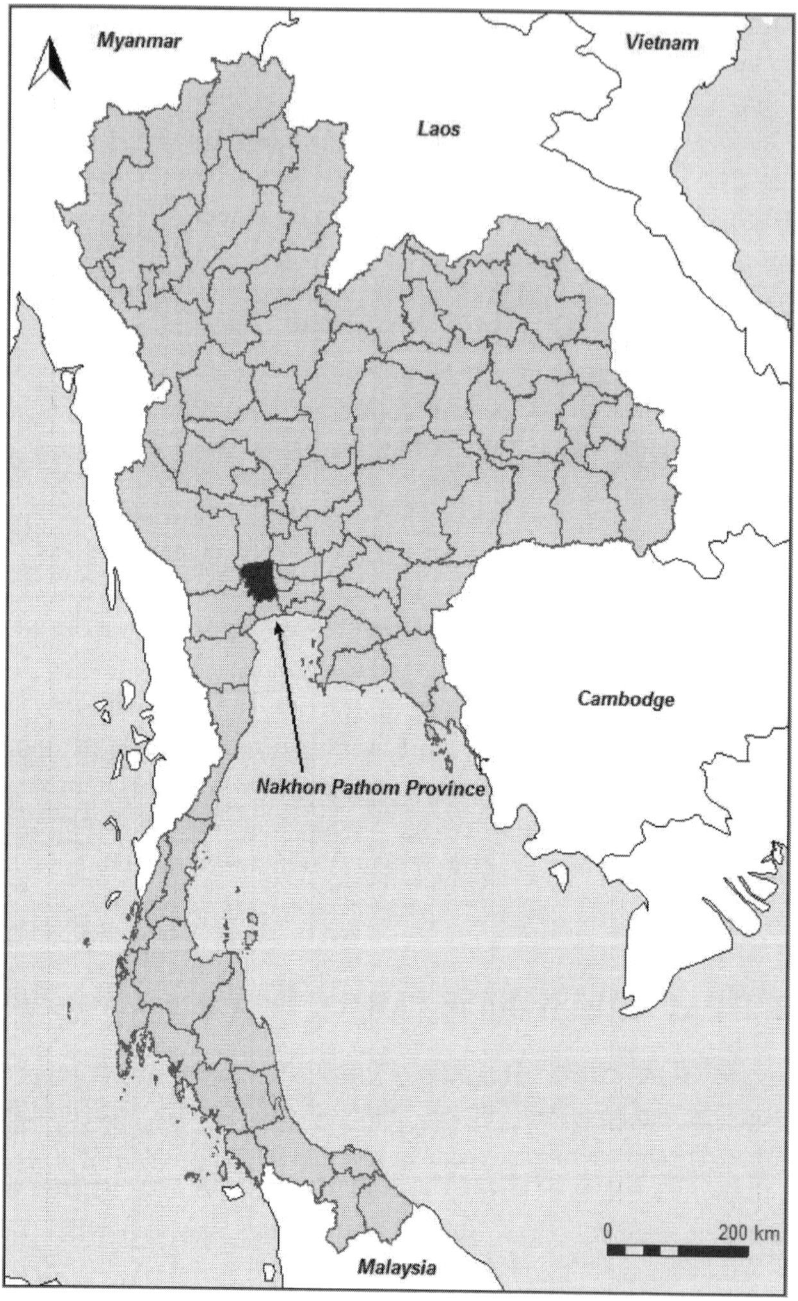


**Figure 1.** Nakhon Pathom province in Thailand.

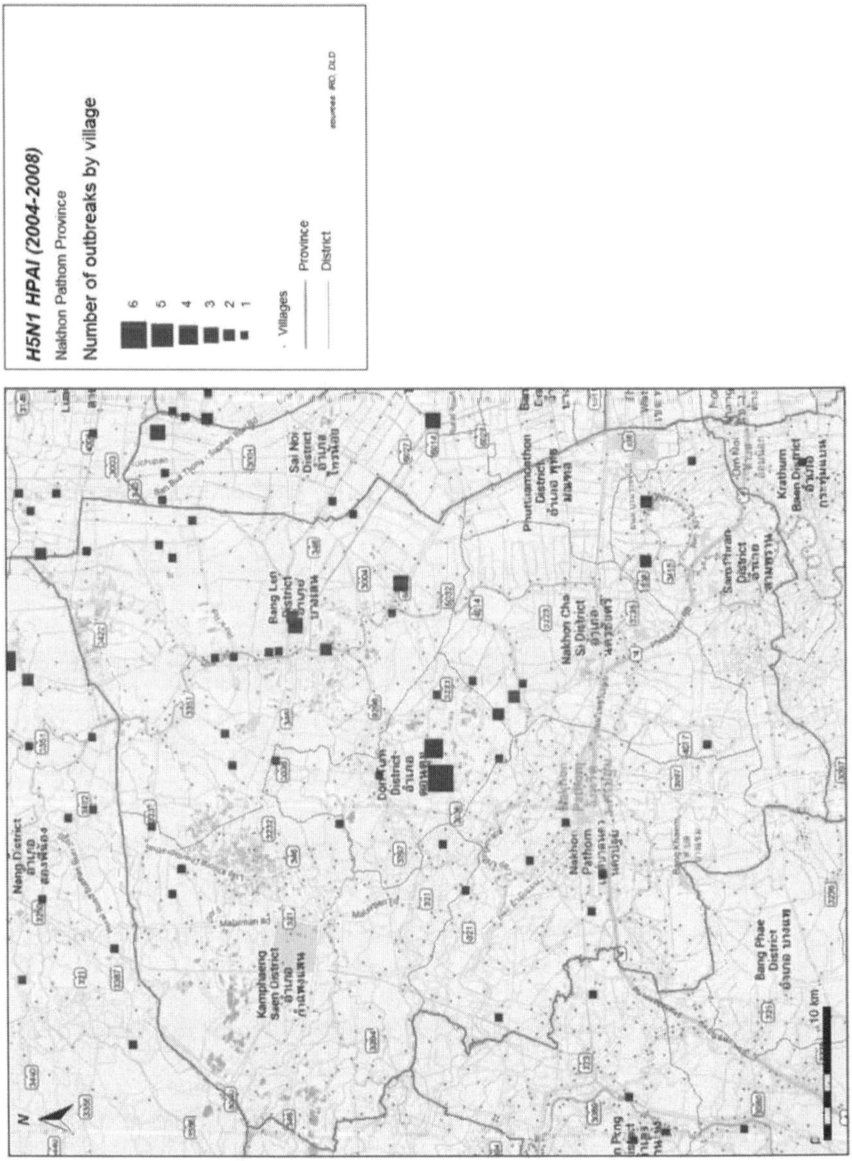


**Figure 2.** H5N1 HPAI outbreaks by village, 2004–2008 in Nakhon Pathom province.

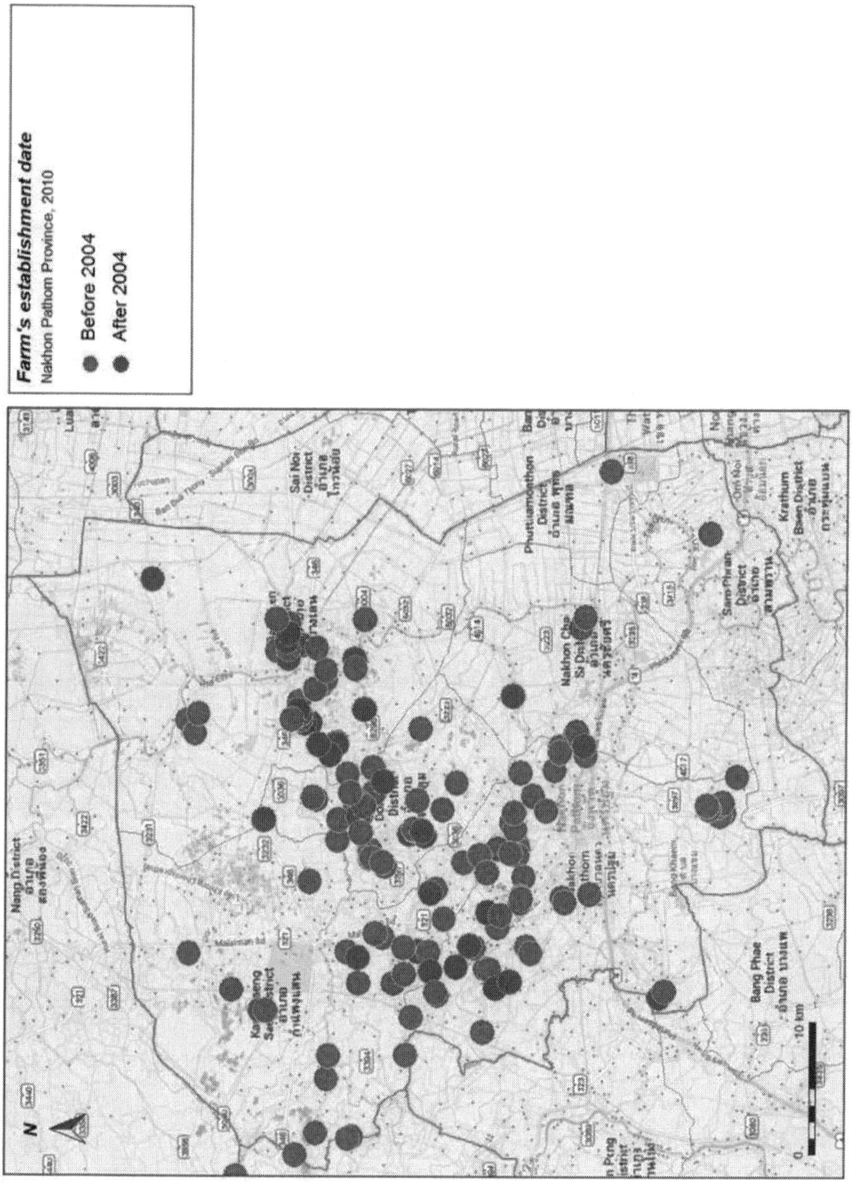


**Figure 3.** Farms in the survey and date of establishment (green: before 2004, brown: after 2004).

## 2.3. Vulnerability Factors, Vulnerability Scores, Statistical and Spatial Analysis

Vulnerability factors in the present study are limited to factors described in DLD, GAP and LFS regulations and are listed in Table 1. All factors are Boolean (yes/no answers). They were grouped into five categories, as per DLD regulations:

- Cat. 1: general condition of the farm, environmental and geographical description, environmental and farm condition, animal health and poultry welfare management;
- Cat. 2: agricultural and agro-commercial practice, slaughtering and selling bird on the farm, to who birds are sold, who organize transports, frequency of transport, "all-in-all-out" practice, restocking time;
- Cat. 3: bio-security measures and farm management, including transports, restocking practice, manure utilization, water management, rodent and insect control, frequency of disinfection, workers routine and hygiene practice, contact between farms;
- Cat. 4: agro-commercial practice of layers farms: egg transportation, disinfection of egg trays, percentages of sale direct to contractors, sale through aggregators;
- Cat. 5: knowledge of the local farmer's regarding HPAI and verification of their commitment to HPAI risk management;

We assigned a weight to each vulnerability factor in order to calculate global vulnerability scores as additive model of vulnerability. Scores are then unique values affected to each farm and make vulnerability evaluation more synthetic.

A vulnerability score for a farm is calculated as a linear weighted sum, by adding the weight of a factor when this factor is positive (yes answer) for the farm. The scores are scaled (divided by the sum of weights) to fit between 0 (lowest) and 1 (highest):

$$Score(k) = \frac{1}{w.} \sum_i w_i F_i\ (k)$$

$$were\ F_i(k) = 1\ if\ factor\ i\ is\ yes\ for\ the\ farm\ k, and\ 0\ else;$$

$$w_i\ is\ the\ weigth\ for\ factor\ i;\ w. = \sum_i w_i$$

(k) The weights correspond to our appreciation of the relative level of importance between factors in term of vulnerability, as experts in this field. As always when using scores based on weights, we assume that our appreciation is subjective and that significant variation in the weighting may occurs if given by different experts.

For each farm, several scores were calculated:

- A score for each category (management and environmental conditions, agricultural practices, bio-security measures, security measures related to egg production, knowledge of the farmer);
- An overall score of vulnerability considering all variables but eliminating variables related to egg production in layer and meat farms;
- An overall score of vulnerability considering all variables including variables related to egg production. Although variables associated with the eggs production are vulnerability factors for the farms producing eggs and are not relevant for farms producing birds, they were included in the calculation of the overall indicator of vulnerability: egg production increases the potential vulnerability with more potentially "at risk" practices and behaviors.

Statistical results by factor were calculated (proportion of positive answers among farms and standard deviation). Multivariate analysis was performed to find relationships between factors (correlations analysis, principal component analysis). Exposure analysis was also performed to find significative differences of vulnerability related to poultry type (duck or chicken) or poultry production (meat or egg) (null hypothesis: no vulnerability difference can be detected between groups; alternative hypothesis: a difference can be detected between groups). We class the scores in two categories, not vulnerable (Score < 0.5) and vulnerable (Score ≥ 0.5) to perform exposure analysis, considering the type (poultry type, production type) as risk factor. We calculated *p*-values for Student *t*-test, odds-ratios and adjusted odds-ratios (confidence intervals calculated at 5% type I error), and *p*-values for Chi-square, Mantel-Haenszel Chi-square and Fisher-Snedecor tests [16].

Spatial analysis of scores was also performed in order to highlight any relationship of vulnerability with environmental condition or geographic organization of space, and to find any spatial pattern of vulnerability susceptible to increase the risk of HPAI farm-to-farm transmission. We performed global spatial autocorrelation using Moran's I index (relationship between value and distance between farms), local spatial association analysis using Getis-Ord LISA index (relationship between the value of a farm and the value of neighbor farms) and search of spatial clusters using spatial scan method [17,18,19]. Data were mapped, statistically and spatially analyzed with the SavGIS geographic information system software [20].

## 2.4. Ethics Statement

In this study, no samples (biological or not) were collected. Farm environmental description and interviews were always conducted anonymously outside the farms. No participant names were recorded, and data were analyzed anonymously. Ethics approval was therefore deemed unnecessary by the Mahidol University IRB.

**Table 1.** Collected factors of vulnerability and related category and weight.

| n° | Factors of Vulnerability | Category * | Weight ** |
|---|---|---|---|
| 1 | Distance to main road less than 50 m | 1 | 7 |
| 2 | No fence at all around the farm | 1 | 9 |
| 3 | Easy entry to the farm | 1 | 8 |
| 4 | A pond is on the farm or next to the farm | 1 | 6 |
| 5 | Open air house on fish pond | 1 | 6 |
| 6 | Dead fowl used as animal food | 2 | 6 |
| 7 | Sale of dead birds | 2 | 9 |
| 8 | Carcass of dead birds throw away | 2 | 7 |
| 9 | Domestic cats and dogs living on the farm | 1 | 4 |
| 10 | Stray dogs/cats on farm | 1 | 8 |
| 11 | Backyard poultry on the farm | 1 | 9 |
| 12 | Presence of other birds | 1 | 5 |
| 13 | Fighting cocks on the farm | 1 | 4 |
| 14 | Presence of exotic birds in cage | 1 | 2 |
| 15 | Presence of other farm animals (pigs, cow, horse, rabbit, boar) | 2 | 4 |
| 16 | Frog breeding on the farm | 2 | 2 |
| 17 | No rodent control | 3 | 6 |
| 18 | No use of insecticides | 3 | 4 |
| 19 | Outsider are permitted to enter the farm | 3 | 9 |
| 20 | Frequent entry/exit from farm more than two times per day | 3 | 9 |
| 21 | Workers do not have working clothing | 3 | 9 |
| 22 | Same employees work on several units on the same farm | 3 | 5 |
| 23 | No farm disinfection | 3 | 9 |
| 24 | Disinfectant barrier at farm entrance does not exist | 3 | 9 |
| 25 | Footbath next to poultry house does not exist | 3 | 8 |
| 26 | Vehicles are not disinfected before entering | 3 | 9 |
| 27 | Dealers/Drivers free movement without changing clothes | 3 | 9 |
| 28 | Poultry houses are not disinfected at end of production | 3 | 9 |
| 29 | Different age groups in same building (no all in/all out) | 2 | 9 |
| 30 | Pond water used for animal drinking | 1 | 7 |
| 31 | Manure not properly stored | 3 | 5 |
| 32 | Manure used as animal feed | 2 | 3 |
| 33 | Use of wet and fresh litter | 2 | 7 |
| 34 | Not daily cleaning of rearing equipment | 3 | 4 |
| 35 | Cleaning of equipment only at the end of production cycle | 3 | 6 |
| 36 | Food delivered by different suppliers | 2 | 7 |
| 37 | Easy access to food storage for rodents and wild birds | 1 | 8 |
| 38 | Poultry sold to trader | 2 | 9 |
| 39 | Restocking before 14 days gap | 3 | 9 |
| 40 | Slaughtering on the farm without proper hygiene facilities | 3 | 3 |
| 41 | Egg trays not disinfected at the farm | 4 | 6 |
| 42 | Eggs are sold on the farm to villagers | 4 | 7 |
| 43 | Eggs sold to different dealers | 4 | 9 |
| 44 | Farmers are not informed about AI | 5 | 9 |
| 45 | Farmers are not concerned about AI as dangerous disease | 5 | 9 |

Notes: * Categories: 1: Farm General Conditions; 2: Agricultural & agro-commercial practices; 3: Bio-security measures and farm management; 4: Layers farms Agro-commercial practices; 5: AI farmer knowledge. ** Weight assigned to each factor to calculate farm vulnerability scores by linear combination.

# 3. RESULTS AND DISCUSSION

## 3.1. Results

All factors were recorded for the 173 farms in the survey, without any missing data. The consent to participate in the door-to-door interview was 100%, and no specific problems occur in the process of data collection (a pilot survey was first carried out in 2010 to adjust the questionnaire used to describe agricultural practices).

Statistical results by factors (proportion of yes answers, standard deviation) are given in Table 2. Scores of vulnerability are presented in Table 3. Category 3 (bio-security measures and farm management) is the one with the worst average score.

**Table 2.** Results of the survey by factor: proportion of yes answers, standard deviation, results by type of production (meat or layer) or type of bird (chicken or duck).

| n° | Vulnerability factors | All farms | Sdt. Dev. | Meat farms | Layer farms | Chicken farms | Duck farms |
|---|---|---|---|---|---|---|---|
| 1 | Distance to main road less than 50 m | 0.18 | 0.38 | 0.16 | 0.22 | 0.18 | 0.18 |
| 2 | No fence at all around the farm | 0.03 | 0.18 | 0.04 | 0.02 | 0.05 | 0.00 |
| 3 | Easy entry to the farm | 0.76 | 0.43 | 0.74 | 0.82 | 0.73 | 0.86 |
| 4 | A pond is on the farm or next to the farm | 0.75 | 0.43 | 0.68 | 0.92 | 0.69 | 0.93 |
| 5 | Open air house on fish pond | 0.11 | 0.31 | 0.12 | 0.06 | 0.14 | 0.02 |
| 6 | Dead fowl used as animal food | 0.53 | 0.50 | 0.50 | 0.57 | 0.50 | 0.59 |
| 7 | Sale of dead birds | 0.01 | 0.11 | 0.01 | 0.00 | 0.02 | 0.00 |
| 8 | Carcass of dead birds throw away | 0.08 | 0.27 | 0.08 | 0.08 | 0.09 | 0.07 |
| 9 | Domestic cats and dogs living on the farm | 0.84 | 0.37 | 0.82 | 0.88 | 0.89 | 0.68 |
| 10 | Stray dogs/cats on farm | 0.27 | 0.45 | 0.32 | 0.14 | 0.26 | 0.32 |
| 11 | Backyard poultry on the farm | 0.18 | 0.38 | 0.15 | 0.26 | 0.16 | 0.25 |
| 12 | Presence of other birds | 1.00 | 0.00 | 1.00 | 1.00 | 1.00 | 1.00 |
| 13 | Fighting cocks on the farm | 0.05 | 0.22 | 0.06 | 0.02 | 0.05 | 0.07 |
| 14 | Presence of exotic birds in cage | 0.04 | 0.20 | 0.05 | 0.02 | 0.04 | 0.05 |
| 15 | Presence of other farm animals (pigs, cow, horse, rabbit, boar) | 0.26 | 0.44 | 0.26 | 0.24 | 0.29 | 0.18 |
| 16 | Frog breeding on the farm | 0.05 | 0.210 | 0.06 | 0.02 | 0.06 | 0.00 |
| 17 | No rodent control | 0.14 | 0.35 | 0.10 | 0.26 | 0.12 | 0.20 |
| 18 | No use of insecticides | 0.94 | 0.23 | 0.95 | 0.89 | 0.94 | 0.95 |
| 19 | Outsider are permitted to enter the farm | 0.37 | 0.48 | 0.31 | 0.51 | 0.34 | 0.45 |

| n° | Vulnerability factors | All farms | Sdt. Dev. | Meat farms | Layer farms | Chicken farms | Duck farms |
|---|---|---|---|---|---|---|---|
| 20 | Frequent entry/exit from farm more than two times per day | 0.67 | 0.47 | 0.60 | 0.84 | 0.66 | 0.70 |
| 21 | Workers do not have working clothing | 0.93 | 0.25 | 0.94 | 0.90 | 0.92 | 0.95 |
| 22 | Same employees work on several units on the same farm | 0.38 | 0.49 | 0.48 | 0.14 | 0.38 | 0.39 |
| 23 | No farm disinfection | 0.41 | 0.49 | 0.37 | 0.51 | 0.38 | 0.50 |
| 24 | Disinfectant barrier at farm entrance does not exist | 0.94 | 0.24 | 0.93 | 0.96 | 0.91 | 1.00 |
| 25 | Footbath next to poultry house does not exist | 0.97 | 0.17 | 0.98 | 0.96 | 0.96 | 1.00 |
| 26 | Vehicles are not disinfected before entering | 0.55 | 0.50 | 0.50 | 0.67 | 0.51 | 0.66 |
| 27 | Dealers/Drivers free movement without changing clothes | 1.00 | 0.00 | 1.00 | 1.00 | 1.00 | 1.00 |
| 28 | Poultry houses are not disinfected at end of production | 0.02 | 0.13 | 0.01 | 0.04 | 0.01 | 0.05 |
| 29 | Different age groups in same building (no all in/all out) | 0.02 | 0.13 | 0.01 | 0.04 | 0.01 | 0.05 |
| 30 | Pond water used for animal drinking | 0.04 | 0.20 | 0.01 | 0.12 | 0.00 | 0.16 |
| 31 | Manure not properly stored | 0.03 | 0.18 | 0.02 | 0.08 | 0.02 | 0.07 |
| 32 | Manure used as animal feed | 0.57 | 0.50 | 0.56 | 0.61 | 0.70 | 0.23 |
| 33 | Use of wet and fresh litter | 0.29 | 0.46 | 0.15 | 0.67 | 0.31 | 0.25 |
| 34 | Not daily cleaning of rearing equipment | 0.27 | 0.44 | 0.27 | 0.26 | 0.34 | 0.05 |
| 35 | Cleaning of equipment only at the end of production cycle | 0.27 | 0.44 | 0.27 | 0.26 | 0.34 | 0.05 |
| 36 | Food delivered by different suppliers | 0.07 | 0.25 | 0.07 | 0.06 | 0.07 | 0.07 |
| 37 | Easy access to food storage for rodents and wild birds | 0.83 | 0.38 | 0.86 | 0.73 | 0.78 | 0.95 |
| 38 | Poultry sold to trader | 0.39 | 0.49 | 0.35 | 0.49 | 0.39 | 0.39 |
| 39 | Restocking before 14 days gap | 0.22 | 0.42 | 0.25 | 0.16 | 0.27 | 0.09 |
| 40 | Slaughtering on the farm without proper hygiene facilities | 0.04 | 0.20 | 0.06 | 0.00 | 0.05 | 0.00 |
| 41 | Egg trays not disinfected at the farm | 0.10 | 0.31 | 0.00 | 0.37 | 0.05 | 0.27 |
| 42 | Eggs are sold on the farm to villagers | 0.21 | 0.41 | 0.00 | 0.73 | 0.19 | 0.27 |
| 43 | Eggs sold to different dealers | 0.15 | 0.36 | 0.00 | 0.51 | 0.14 | 0.18 |
| 44 | Farmers are not informed about AI | 0.01 | 0.13 | 0.01 | 0.02 | 0.01 | 0.05 |
| 45 | Farmers are not concerned about AI as dangerous disease | 0.63 | 0.48 | 0.72 | 0.41 | 0.64 | 0.59 |

**Table 3.** Summary statistics for five categories of vulnerability scores of commercial poultry farms in Nakhon Pathom, Thailand, 2011.

| Vulnerability Score Category | Min. | Max. | Mean | Median | Sdt. Dev. |
|---|---|---|---|---|---|
| General characteristics of the farm | 0.11 | 0.66 | 0.39 | 0.37 | 0.11 |
| Agricultural and commercial practices | 0 | 0.54 | 0.20 | 0.19 | 0.12 |
| Biosecurity measures, farm management | 0.25 | 0.79 | 0.52 | 0.52 | 0.10 |
| Layers farms practices | 0 | 1 | 0.16 | 0 | 0.29 |
| AI farmer knowledge | 0 | 0.5 | 0.32 | 0.5 | 0.24 |
| Global score | 0.17 | 0.58 | 0.38 | 0.37 | 0.07 |

Systematic correlation analysis did not reveal strong relationships between vulnerability factors. A multivariate principal component analysis with all the 45 factors of the survey (Table 1) shows a first component explaining only 10% of the total variance.

Analysis of vulnerability by type of production (meat or layer) revealed that excessive vulnerability correlates with farms producing eggs (Table 4). The type of production is very significant for vulnerability since factors associated with eggs production may strongly contribute to increasing vulnerability. Furthermore, the same results are found whenever we take into account the vulnerability factors directly related to egg production.

**Table 4.** Analysis of vulnerability scores by type of production (meat or layer).

| Vulnerability Score Category | Meat * | Layer | *p*-value (T) | OR | $Chi^2$ |
|---|---|---|---|---|---|
| General characteristics of the farm | 0.38 | 0.40 | 0.17 | 1.32 [0.52, 3.31] | 0.34 |
| Agricultural and commercial practices | 0.18 | 0.27 | 0.00005 | 3.77 [1.63, 8.69] | 10 |
| Biosecurity measures, farm management | 0.51 | 0.55 | 0.001 | 5.64 [2.27, 14.01] | 16 |
| AI farmer knowledge | 0.37 | 0.21 | 0.00006 | 0.27 [0.14, 0.54] | 14 |
| Global score | 0.36 | 0.43 | <0.000001 | 13.24 [4.12, 42.51] | 27 |
| Score w/t cat. 4 | 0.39 | 0.42 | 0.0006 | 5.53 [2.29, 13.32] | 16 |

Notes: * Reference group. T. Student. $Chi^2$: Chi square value.

Analysis of vulnerability by poultry type (duck or chicken) shows no significant differences (Table 5). With the overall score classified into two classes (not vulnerable, $s < 0.5$ and vulnerable, $s \geq 0.5$), the odd ratio adjusted by poultry production doesn't show a significant difference between classes.

Finally, the study of scores using all possible combinations between production type (meat or layer) and poultry type (duck or chicken) shows that farms producing duck eggs are the most vulnerable in all categories, with a difference between groups always significant at a 5% type I error (Table 6).

**Table 5.** Analysis of vulnerability scores by poultry type (chicken or duck), with OR adjusted by production type (meat or layer), and Mantel-Haenszel Chi².

| Vulnerability Score Category | Chicken * | Duck | *p*-value (T) | OR (Ajusted) | Chi² (MH) |
|---|---|---|---|---|---|
| General characteristics of the farm | 0.37 | 0.43 | 0.0028 | 1.25 [0.46, 3.43] | 0.15 |
| Agricultural and commercial practices | 0.21 | 0.18 | 0.056 | 1.35 [0.55, 3.32] | 0.31 |
| Biosecurity measures, farm management | 0.51 | 0.53 | 0.19 | 0.81 [0.27, 2.44] | 0.22 |
| Layers farms practices | 0.13 | 0.23 | 0.006 | 0.93 [0.28, 3.10] | 0.01 |
| AI farmer knowledge | 0.32 | 0.32 | 0.42 | 1.09 [0.51, 2.32] | 0.05 |
| Global score | 0.38 | 0.40 | 0.03 | 1.41 [0.65, 3.06] | 0.81 |
| Score w/t cat. 4 (Layers farms practices) | 0.39 | 0.41 | 0.08 | 2.28 [0.89, 5.86] | 3.06 |

Notes: * Reference group. T: Student. Chi² (MH): Mantel-Haenszel Chi square value.

**Table 6.** Analysis by poultry type and production type.

| Vulnerability Score | All | Meat | | Layer | | *p*-value (F *) |
|---|---|---|---|---|---|---|
| | | Chicken | Duck | Chicken | Duck | |
| General characteristics of the farm | 0.39 | 0.37 | 0.42 | 0.38 | 0.44 | 0.046 |
| Agricultural and commercial practices | 0.20 | 0.19 | 0.15 | 0.28 | 0.24 | 0.00003 |
| Biosecurity measures, farm management | 0.52 | 0.51 | 0.50 | 0.53 | 0.58 | 0.02 |
| Layers farms practices | 0.16 | 0.00 | 0.00 | 0.50 | 0.62 | <0.00001 |
| AI farmer knowledge | 0.32 | 0.36 | 0.37 | 0.21 | 0.22 | 0.002 |
| Global score | 0.38 | 0.36 | 0.36 | 0.42 | 0.46 | <0.00001 |
| Score w/t cat.4 (Layers farms practices) | 0.40 | 0.39 | 0.39 | 0.41 | 0.44 | 0.02 |

Note: * F: Fisher-Snedecor.

Spatial analysis was performed to help in identifying further contributing factors in environment or geographic context or organization of space. Figure 4 and Figure 5 show the spatial distribution of the vulnerability score for layer farms and for meat farms. Significant global spatial autocorrelation of global vulnerability score (Z-score > 1.96) is found within a short radius for the spatial weight used in the spatial autocorrelation index (search radius up to 10 km). With automatic cluster detection using a spatial statistic scan, we found a most significant spatial cluster for the vulnerability score in our study, and this cluster is spatially close to the most significant cluster of the H5N1 HPAI cases that occurred during the epidemics from 2004 to 2008, and automatically detected with the same technique (Figure 6). Local spatial association is also found (Figure 7), showing again a clustered pattern, especially for high level of vulnerability.

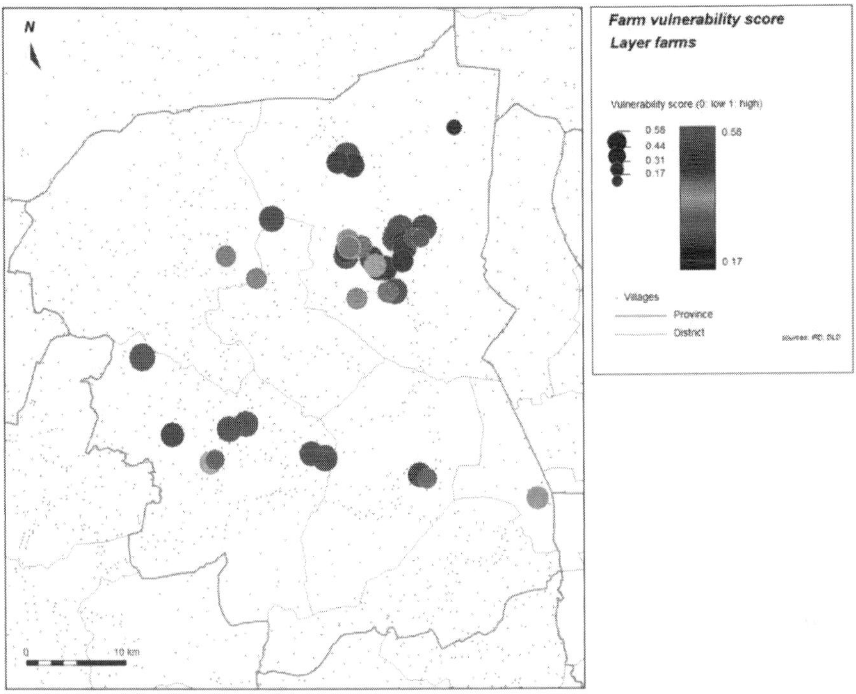


**Figure 4.** Score of vulnerability, layer farms sector 2 and 3, Nakhon Pathom province, Thailand. Level of vulnerability is given by color graduation and symbol width.

Spatial analysis performed by separating farms into two groups by poultry production (meat or layer) showed no significant spatial clustering in each group separately.

## 3.2. Discussion

For more than 20 years, the poultry production in Thailand has radically shifted to a more industrial structure, and in 2002 more than 70% of national poultry production was oriented to the export market [13]. Industrial poultry production system in Thailand consists of vertically integrated companies controlling every stage of the production chain, from breeding to marketing processed poultry [21]. Those companies offer contractual agreements to private farms and purchase poultry products from them. The firm holds considerable control over raw material production without ownership of the production units. Contract farmers have an assured market for their poultry, access to the firm's services, and easier access to credit from banks [22]. Independent farmers mostly sell their larger surpluses to aggregators who drive pickups from village to village, purchasing birds from numerous different farms. A small amount of eggs and live or slaughter birds are sold to "End Users" (neighbors, restaurants or village, market vendors). Due to this production system, 98% of commercial poultry producers (sector 1–3) belongs to sector 2 and 3 in Thailand [3]. Sector 4 still exists as a source of suppliers for local markets.

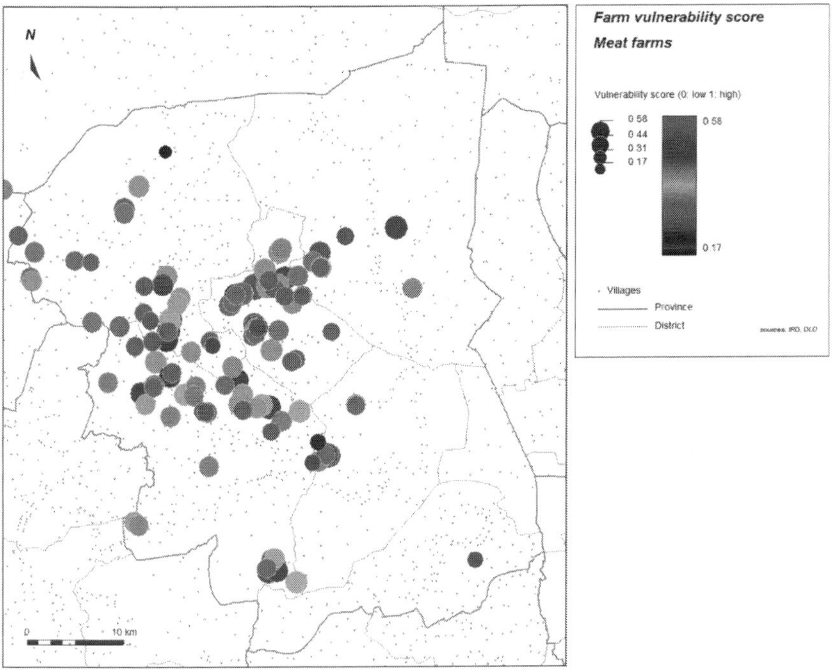


**Figure 5.** Score of vulnerability, meat farms sector 2 and 3, Nakhon Pathom province, Thailand. Level of vulnerability is given by color graduation and symbol width.

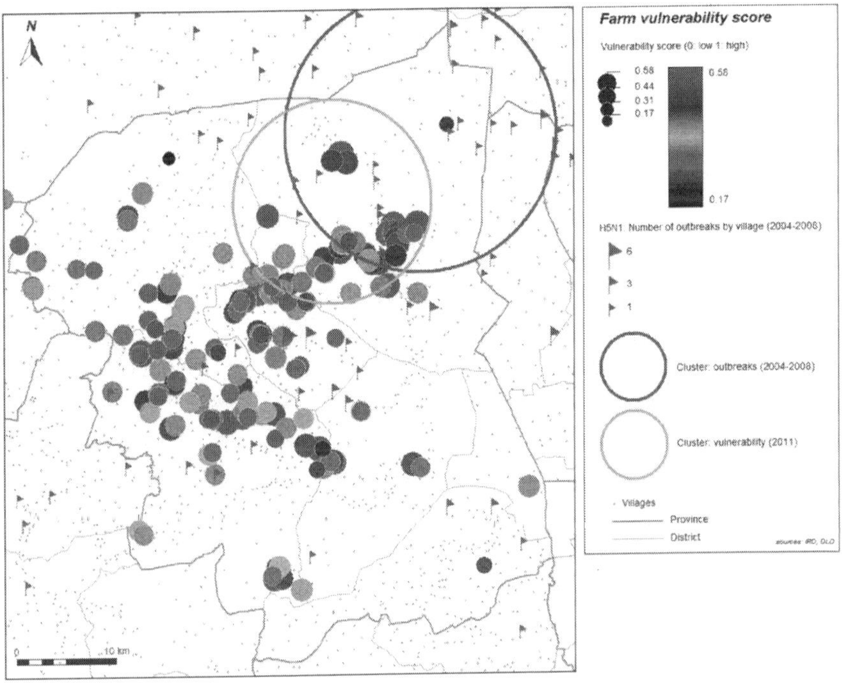


**Figure 6.** Most significant cluster of vulnerability (yellow), farms sector 2 and 3, and most significant cluster (red) of 2004–2008 H5N1 HPAI outbreaks, Nakhon Pathom province, Thailand.

Our goal was therefore to determine the actual level of vulnerability of this medium and small scale commercial poultry production system (sector 2 and 3) and to determine the most hazardous poultry farming practices related to the risk of avian influenza contamination and spread.

Our study clearly showed that vulnerability among farms in the province of Nakhon Pathom is still very high. It appears that the vulnerabilities related to poor farm management are predominant. Major deficiencies in bio-security measures were found in the areas of farm management, including transport, restocking practice, manure utilization, water management, rodent and insect control, frequency of disinfection, worker routines, hygiene practices and contact between farms. This is undoubtedly the result of a lack of awareness among farmers regarding the potential risks of epidemics. They are relatively complying with current regulations, but not all are entirely convinced that the risk of infections is still real, after a long period without warning since the last outbreak (in 2008).

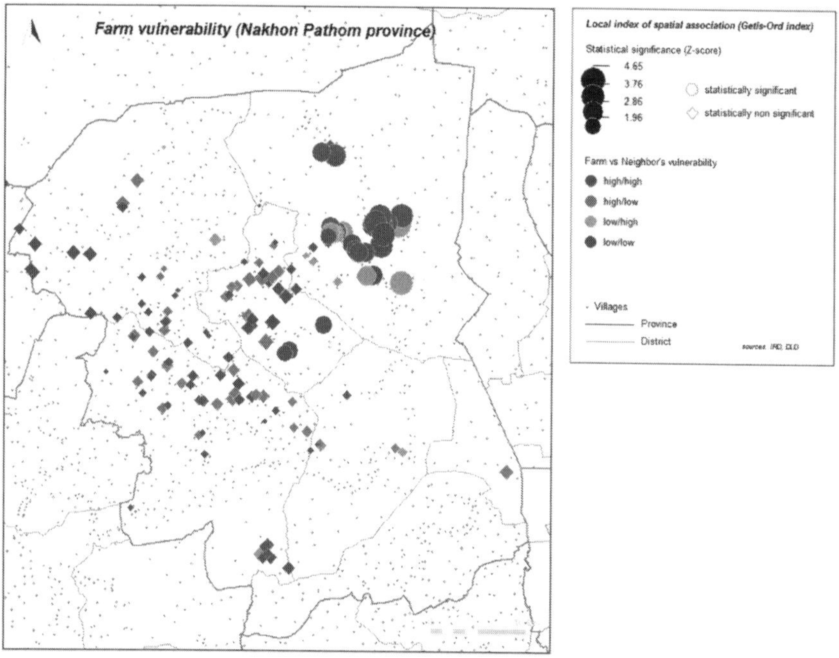


**Figure 7.** Local index of spatial association (Getis-Ord) for vulnerability, farms sector 2 and 3, Nakhon Pathom province, Thailand.

Easy access to farm through entrances was found in 76% of farms, as the doors are open with no locks or warning sign. This percentage is very high and this is a major issue in vulnerability regarding farm to farm contamination.

75% of farms are located next to a pond or a pond is present at the farm. 19% of farms practice an integrated poultry and fish farming. Droppings and food spilled by chickens fall directly into the water and feed the fish. If birds are housed without proof housing this practice is considered as high risk to disseminate avian influenza virus [6,23,24,25,26]. Furthermore, constant presence of wild birds or ducks attracted by the pond can contribute to the spreading of the disease. Finally, disinfection of poultry houses is not practiced in integrated poultry-fish farms since farmers are afraid of killing the fish. All farmers declare the regular presence of wild birds on the farm. 82% have no proper food storage or keep an open door at poultry food storage.

Although 85% of farmers replied positively to the use of rodent control, 83% of them use cats as the unique means of rodent control. The use of insecticides is very uncommon, and 94% think insects are not an issue. Even if insects themselves do not represent a known risk for the spread of avian influenza, insect infestation is a marker of poor hygiene and poor maintenance of the farm.

The transport of poultry, food, or fomite is an important risk factor for the transmission and dispersion of avian influenza [27,28]. On contractor farms, the company tracks and collects poultry or delivers food directly to the farm, unlike 39% of independent farms who delivers their products to various aggregators. These aggregators move from one farm to another, thus increasing the risk of avian influenza virus entry [29]. This practice represents a significant risk of spreading pathogens between farms if the disinfections of trucks are not applied at the entries and exits of the farm [30]. Similarly, the transport of poultry to the slaughterhouse several times per production cycle and restocking with non-compliance the rule "all in, all out" is also a risk factor [6]. In 93% of visited farms, disinfectant barriers at the farm entrance do not exist. 55% of farms do not practice trucks disinfection. Moreover, at all farms, truck drivers leave vehicles and walk around in their own clothes. These results show a very high vulnerability issue.

Employers of 93% of the farms do not provide or request work clothing, and at 38% of farms the same employees work on several different poultry units. Most farms do not provide food for the workers, and they frequently go out (more than two times a day) to buy a meal. Moreover, 37% of farms permit street food vendors to enter the farms and sell food to workers. Again, these results show a high vulnerability issue regarding virus dissemination.

Regular cleaning of poultry rearing equipment is practiced on only 26% of farms. Farm disinfection is applied at 59%, and no one in a poultry farm above a fishpond uses disinfectant. Disinfection of poultry houses at the end of the poultry production cycle is widely applied but still not applicable on poultry rearing equipment and egg trays. As animal food, fresh manure is used at 57% of farms, and 50% use dead fowl of farms as animal food.

Farms producing eggs are more vulnerable than farms producing birds and this result is consistent with the type of farms mostly affected during the 2004–2008 outbreaks [2]. Layer farms are more vulnerable than meat farms: factors associated with egg production strongly contributes to the vulnerability, and many farms producing eggs have low results regarding vulnerability factors linked to egg production.

Although almost all farmers knew about avian influenza, 63% of them do not think it is a dangerous disease and therefore are not concerned. Farmers adhere only to what is requested by the DLD GAP regulations. The long period without warning since the last outbreak of HPAI could also be an element of a relative lessening of farmers' vigilance.

Spatial clustering of vulnerability is also an important bio-security issue; infection in a farm may spread to neighboring farms if vulnerability of neighbors is high. Compartmentalization (defined primarily by management and husbandry practices related to biosecurity) is recommended by FAO [6,23] and has been promoted to limit disease spread between farms. Nevertheless, our results show that many vulnerability issues are related to contacts (*i.e.*, easy access to farms, behaviors of workers, low disinfection at entrance) and are clustered. These spatial clusters of vulnerability need to receive special attention from authorities to respect the purpose of compartmentalization.

## 4. CONCLUSIONS

The results of our survey in the Nakhon Pathom province has shown that numerous vulnerabilities still exist and could represent a significant risk for large dissemination in the event of HPAI re-emergence. These results strongly suggest that bio-security practices on the majority of farms still need improvement. Particular attention should be conducted when vulnerability is spatially clustered.

It was also noted that category 3 (bio-security measures and farm management) is the one with the worst average score. In order to improve on-farm bio-security, it is necessary to enhance the capacity of poultry workers and farm owners to recognize the importance of these security measures. This can only be achieved through careful communication with an engagement of farmers and communities. DLD GAP regulations should be sharpened; stricter LFS regulations and GAP formulation with the implication of details is highly recommended for bio-security measures and farm management with a special attention for egg production management and practices.

## ACKNOWLEDGMENTS

We acknowledge AIRD (Agence Inter-établissements de Recherche pour le Développement) for funding this project under the ARDIGRIP project.

## AUTHOR CONTRIBUTIONS

Marc Souris and Dubravka Selenic conceived and designed the study, and performed interpretation; Marc Souris, Dubravka Selenic, Supaluk Khaklang and Suwannapa NInphanomchai prepared the draft of the manuscript; Dubravka Selenic, Supaluk Khaklang and uwannapa NInphanomchai performed field work and interviews; Marc Souris performed statistical and spatial analysis; Marc Souris and Jean-Paul Gonzalez provided intensive review of the manuscript; Pattamaporn Kittayapong review the manuscript and gave a final approval for the manuscript to be submitted for publication. All authors edited and commented on the manuscript.

## CONFLICTS OF INTEREST

The authors declare no conflict of interest.

## REFERENCES

1 OIE. Update on Highly Pathogenic Avian Influenza in Animals (Type H5 and H7). 2013. Available online: http://www.oie.int/en/animal-health-in-the-world/update-on-avian-influenza (accessed on 24 December 2013).

2 Souris, M.; Gonzalez, J.P.; Shanmugasundaram, J.; Corvest, V.; Kittayapong, P. Retrospective space-time analysis of H5N1 Avian Influenza emergence in Thailand. *Int. J. Health Geogr.* **2010**, *9*.

3 Rushton, J.; Viscarra, R.; Bleich, E.G.; McLeod, A. Impact of Avian Influenza Outbreaks in the Poultry Sectors of Five South East Asian Countries (Cambodia, Indonesia, Lao PDR, Thailand, Vietnam) Outbreak Costs, Responses and Potential Long Term Control. In *TCP/RAS/3010*; FAO: Rome, Italy, 2007; Volume 10, pp. 1–25.

4 Beaudoin, A.L.; Kitikoon, P.; Schreiner, P.J.; Singer, R.S.; Sasipreeyajan, J.; Amonsin, A.; Gramer, M.R.; Pakinsee, S.; Bender, J.B. Risk factors for exposure to influenza a viruses, including subtype H5 viruses, in Thai free-grazing ducks.*Transbound. Emerg. Dis.* **2012**.

5 Costales, A. Thailand Poultry Sector Brief. In *Food and Agriculture Organization*; PPLPF: Rome, Italy, 2002.

6 *Recommendations on the Prevention, Control and Eradication of Highly Pathogenic Avian Influenza (HPAI) in Asia*; FAO: Rome, Italy, 2004.

7 Tiensin, T.; Chaitaweesub, P.; Songserm, T.; Chaisingh, A.; Hoonsuwan, W.; Buranathai, C.; Parakamawongsa, T.; Premashthira, S.; Amonsin, A.; Gilbert, M.; *et al.* Highly

pathogenic avian influenza H5N1, Thailand, 2004. *Emerg. Infect. Dis.* **2005**, *11*, 1664–1672.

8 Gilbert, M.; Chaitaweesub, P.; Parakamawongsa, T.; Premashthira, S.; Tiensin, T.; Kalpravidh, W.; Wagner, H.; Slingenbergh, J. Free-grazing ducks and highly pathogenic avian influenza, Thailand. *Emerg. Infect. Dis.* **2006**, *12*, 227–234.

9 Paul, M.; Wongnarkpet, S.; Gasqui, P.; Poolkhet, C.; Thongratsakul, S.; Ducrot, C.; Roger, F. Risk factors for highly pathogenic avian influenza (HPAI) H5N1 infection in backyard chicken farms, Thailand. *Acta Trop.* **2011**, *118*, 209–216.

10 Keawcharoen, J.; Broek, J.; Bouma, A.; Tiensin, T.; Osterhaus, A.; Heesterbeek, H. Wild birds and increased transmission of highly pathogenic avian influenza (H5N1) among poultry, Thailand. *Emerg. Infect. Dis.* **2011**, *17*, 1016–1022.

11 Tiensin, T.; Nielen, M.; Vernooij, H.; Songserm, T.; Kalpravidh, W.; Chotiprasatintara, S.; Chaisingh, A.; Wongkasemjit, S.; Chanachai, K.; Thanapongtham, W.; *et al.* Transmission of the highly pathogenic avian influenza virus H5N1 within Flocks during the 2004 epidemic in Thailand. *J. Infect. Dis.* **2007**, *196*, 1679–1684.

12 *Avian Influenza Situations and Control Measures*; Department of Livestock Development (DLD): Bangkok, Thailand, 2004.

13 Ranong, V.N. Structural changes in Thailand's poultry sector: avian influenza and its aftermath. *TDRI Quart. Rev.***2008**, *23*, 3–10.

14 Wei, H.; Aengwanich, W. Biosecurity evaluation of Poultry Production Clusters (PPCs) in Thailand. *Int. J. Poult. Sci.***2012**, *11*, 582–588.

15 Chantong, W.; Kaneene, J.B. Poultry raising systems and highly pathogenic avian influenza outbreaks in Thailand: the situation, associations, and impacts. *Southeast Asia. J. Trop. Med. Public Health* **2011**, *42*, 596–608.

16 Bouyer, G.; Hémon, D.; Cordier, S.; Derriernnic, F.; Stücker, I.; Stengel, B.; Clavel, J. *Epidemiologie, Principles et Methods Quantitative*; Les Editions INSERM: Paris, France, 1995.

17 Pfeiffer, D.; Robinson, T.; Stevenson, M.; Stevens, K.; Rogers, D.; Clements, A. *Spatial Analysis in Epidemiology*; Oxford University Press: Oxford, UK, 2008.

18 Souris, M.; Bichaud, L. Statistical methods for bivariate spatial analysis in marked points. Example in spatial epidemiology. *Spat. Spatio-Temporal Epidemiol.* **2011**, *2*, 227–234.

19 Kulldorff, M. A spatial scan statistic. *Commun. Statist.—Theor. Method.* **1997**, *26*, 1481–1496.

20 SavGIS Geographic Information System. Available online: www.savgis.org (accessed on 24 December 2013).

21 Heft-Neal, S.; Otte, J.; Pupphavessa, W.; Roland-Holst, D.; Sudsawasd, S.; Zilberman, D. Supply Chain Auditing for Poultry Production in Thailand. In *Pro-poor Livestock Policy Initiative Research Report*; FAO: Rome, Italy, 2008.

22 Glover, D.; Kusterer, K. *Small Farmers, Big Business: Contract Farming and Rural Development*; St. Martin's Press: New York, NY, USA, 1990.

23 FAO/OIE/WHO. *Consultation on Avian Influenza and Human Health: Risk Reduction Measures in Producing, Marketing, and Living with Animals in Asia*; FAO: Rome, Italy, 2005.

24 Webster, R.G.; Yakhno, M.; Hinshaw, V.S.; Bean, W.J.; Murti, K.G. Intestinal influenza: Replication and characterization of influenza viruses in ducks. *Virology* **1978**, *84*, 268–278.

25 Utterback, W. Update on Avian Influenza through February 21 1984 in Pennsylvania and Virginia. In Proceedings of the 33rd Western Poultry Disease Conference, University of California, Davis, CA, USA, 27–29 February 1984; pp. 4–7.

26 Feare, C.J. *Fish Farming and the Risk of Spread of Avian Influenza*; WildWings Bird Management, BirdLife International:

Cambridge, UK, 2006; p. 11. Available online: http:// www.naturecanada.ca/pdf/bli_fish_farming_review.pdf(accessed on 24 December 2013).

27 Halvorson, D.A.; Karunakaran, D.; Newman, J.A. Avian influenza in caged laying chickens. *Avian Dis.* **1980**, *24*, 288–294.

28 Glass, S.E.; Naqi, S.A.; Grumbles, L.C. Isolation of avian influenza viruses in Texas. *Avian Dis.* **1981**, *25*, 545–549.

29 Delforge, I. Contract Farming in Thailand: A View from the Farm. In *Occasional Paper 2, Focus on the Global South, CUSRI*; Chulalongkorn University: Bangkok, Thailand, 2007.

30 Thomas, M.E.; Bouma, A.; Ekker, H.M.; Fonken, A.J.; Stegeman, J.A.; Nielen, M. Risk factors for the introduction of high pathogenicity Avian Influenza virus into poultry farms during the epidemic in the Netherlands in 2003. *Prev. Vet. Med.* **2005**, *69*, 1–11.

# CHAPTER 4

# Characterization of Newcastle Disease Virus and poultry-handling practices in live poultry markets, Ethiopia

**Delesa Damena Mulisa[1], Menbere Kidane W/Kiros[1], Redeat Belaineh Alemu[1], Melaku Sombo Keno[1], Alice Furaso[2], Alireza Heidari[1], Tesfaye Rufael Chibsa[1] and Hassen Chaka Chunde[1]**

[1]National Animal Health Diagnostic and Investigation Center, Sebeta, Ethiopia.

[2]Research and Innovation Department, Istituto Zooprofilattico Sperimentale delle Venezie, OIE/FAO and National Reference Laboratory for Newcastle Disease and Avian Influenza,OIE collaborating Center for Diseases at the Human-Animal Interface, Legnaro, Padova, Italy.

## ABSTRACT

Newcastle disease represents the most severe poultry disease responsible for marked economic losses in Ethiopia. To provide a molecular characterization of Newcastle disease viruses circulating in this country, a cross sectional survey was conducted at five selected live poultry market sites in Addis Ababa. In addition, baseline data on the live poultry market system were acquired through a detailed questionnaire submitted to poultry traders. We identified 44/146 positive samples, 29 of which were virulent strains belonging to sub-genotype VIf. The very poor biosecurity practices, which have resulted from responses of the participants, suggest that they might have had a heavy impact in the spread of the disease. This study provides important information on epidemiology and control of NDV in Ethiopia and highlights the importance of implementing surveillances and biosecurity practices in live poultry markets.

## KEYWORDS

Newcastle disease virus, Biosecurity practices, Molecular characterization, Live poultry markets, Ethiopia

## 1. INTRODUCTION

Poultry play an important economic, nutritional and socio-cultural role in the livelihoods of poor rural households in developing countries, including Ethiopia. The total poultry population in Ethiopia is estimated at 43 million, 97% of which are village chickens (Central Statistical Agency 2012). Poultry rearing is particularly important to women, who often own and manage chickens and the resulting income is often used to support education of children. Despite its role in raising incomes and reducing poverty in local communities of Ethiopia, poultry production is hampered by wide arrays of constraints among which infectious diseases, such as: Newcastle disease, Infectious Bursal Disease, Mycoplasmosis, Pasteurellosis and Salmonellosis, are the major ones (Chaka et al. 2012).

Newcastle disease(ND) is one of the major problems in village chickens in most parts of Ethiopia (Nasser 1998; Tadelle and Jobre 2004; Mazengia 2012). The disease has become endemic in poultry population and recurs every year inflicting heavy losses (Tadelle and Jobre2004). The highest rate of ND outbreaks from March to May is suggested to be associated with high rate of chicken marketing for Easter

(Spradbrow 1999). The main movement of chicken marketing is from periphery to the center (rural to towns) which favors the spread of diseases all over the country (Dessei and Ogle 2001). However, there is acute lack of information on bio-security situation and roles of poultry marketing practices in diseases dissemination in the country (Shewantasew et al. 2012).

ND is caused by virulent strains of Newcastle diseases virus (NDV) or avian paramyxovirus type 1 (APMV-1) that belongs to the genus*Avula* and family *Paramyxoviridae* (Mayo 2002). NDV has an enveloped, single-stranded negative sense RNA genome of approximately 15 kb that contains six genes encoding major structural proteins such as: nucleocapsid protein (NP), phosphoprotein(P), matrix protein(M), fusion protein(F), hemagglutinin-neuraminidase(HN), and the RNA- dependent RNA polymerase(L) (Kattenbelt et al. 2006) as well as two additional proteins, V and W from the P gene by a mechanism called RNA editing (Steward et al. 1993).

The key contributors to NDV pathogenicity are the formation of an active fusion protein upon cleavage of the F protein precursor (F0) as well as the presence of a number of basic residues in the fusion protein cleavage site (Toyoda et al. 1987; Glickman et al. 1988).

NDV that are virulent for chickens have a multibasic amino acid sequence $^{112}$R/K-R-Q-K/R-R$^{116}$ at the C–terminus of the F2 protein and F (phenylalanine) at residue 117, which is the N-terminus of the F1 protein, whereas the viruses of low virulence have a monobasic amino acid sequences in the same region of $^{112}$G/E-K/R-Q-G/E-R$^{116}$ and L (leucine) at residue 117 (Kim et al. 2008; Office of International Epizootes2012).

Although ND represents the most severe poultry disease responsible for marked economic losses in Ethiopia, virological and epidemiological information concerning the virus strains circulating in the country especially in village chickens are extremely scarce. The aim of this study was therefore, to detect NDV and to perform a molecular characterization of the virulent strains currently circulating at five live poultry market sites in Addis Ababa. In addition, we investigated the biosecurity practices implemented at the market sites to identify the factors that can contribute to the spread of the virus.

# 2. MATERIALS AND METHODS

## 2.1 Sample origin

The study was conducted in December 2012 in five live poultry market sites in Addis Ababa namely: Kotebe, Merkato, Shola, Saris and Semen (Adisu gebeya) (Figure 1). These are among the biggest poultry market sites in Addis Ababa that host chickens originating from most parts of the country, mainly from places located within 300 kms radius such as: Sodo, Hosaina, Dessie, Shashemane, Jimma, and Ambo. All the markets are outdoor markets where chickens are kept in wire mesh cages, each housing up to 40 birds during the day and night.

## 2.2 Sampling, data collection and virus isolation

A total of 365 chickens were sampled: From each market, 70 birds were sampled except Merkato from where 85 birds were sampled by considering its larger size compared to the others. Thus, 73 tracheal swabs (pool of five) and 73 cloacal (pool of five) were collected. In separate cryovial containing 2 ml of freshly prepared viral transport media (VTM).

The swabs were collected in pairs (Tracheal swab pools and Cloacal swab pools were always collected from the same chickens). After sampling, the specimens were transported to the National Animal Health Diagnostic and Investigation Center (NAHDIC) laboratory and stored at -80°C until processing. Besides, a standardized questionnaire was administered to the poultry traders in each market sites. A total of 100 volunteer traders were interviewed to assess the bio-security practices in live poultry markets. The questionnaire included questions about the source of the birds, means of transportation, the number of birds in each market site, cleaning practices at the market sites including the cages, health status of the birds and disposal of manures and dead chickens.

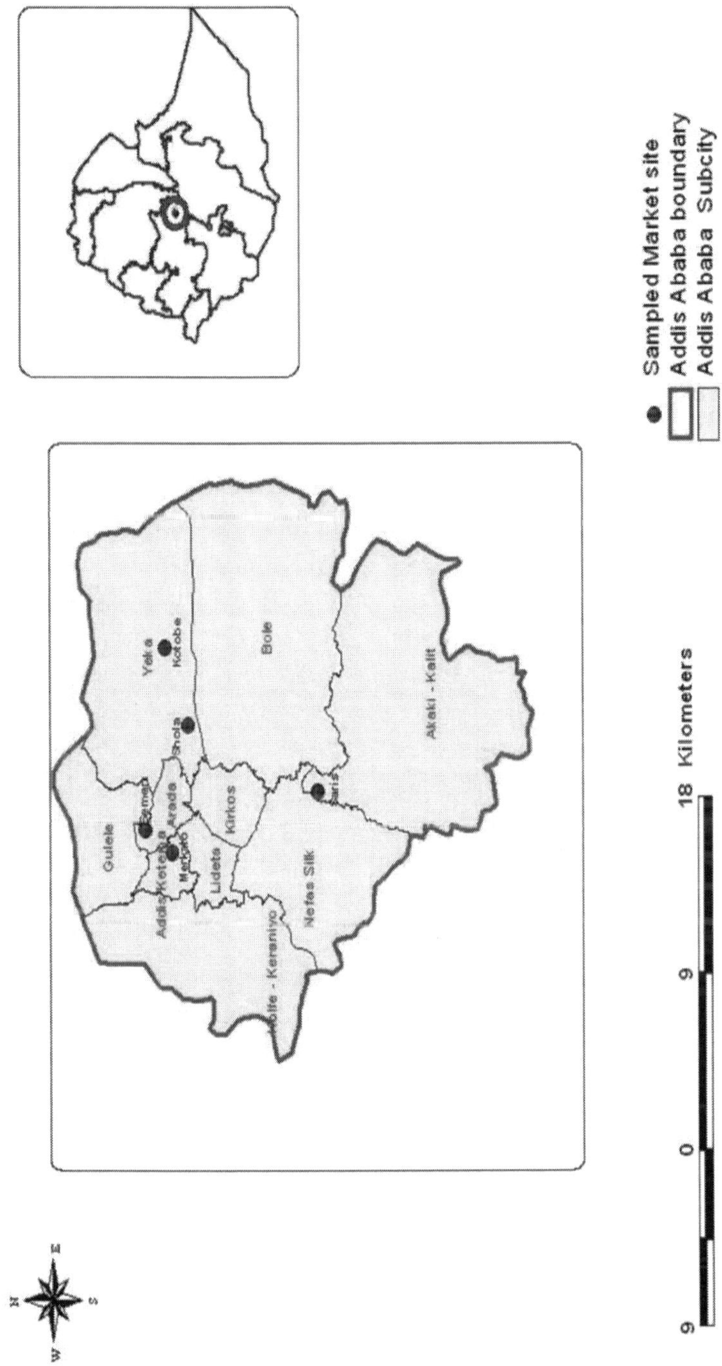


**Figure 1** Map of Study area. Poultry market sites where samples were collected.

Viruses were isolated from the swab samples by standard virus isolation methods in embryonated chicken eggs. Three embryonated chicken eggs of 9 to 11-day-old per sample were used for inoculation (Alexander and Senne 2008). The inoculated eggs were candled every 24 hours to check embryo vitality. Eggs containing dead embryo on each day and those remained at the end of incubation period, were removed from incubator and chilled at +4°C overnight. Allantoic fluids were harvested and tested by Haemagglutination (HA) test for its ability to haemagglutinate chicken RBCs.

## 2.3 Haemagglutination and Haemagglutination Inhibition tests

The HA assay was carried out in microtitre plate as outlined by the World Organization for Animal Health (Office of International Epizootes2012). HA positive samples were tested by Heamaglutination Inhibition (HI) reaction to specifically determine the haemagglutinating agents. Inactivated antigen for HA test and positive and negative sera for HI controls were obtained from Istituto Zooprofilattico Spermentale delle Venezie (OIE/FAO Reference Laboratory for AI and ND), Padua, Italy.

## 2.4 Polymerase Chain Reaction tests

Forty four HA and HI positive allantoic fluids were analyzed by real-time reverse transcription (RT-PCR) tests. Viral RNA extraction from HI positive allantoic fluid was conducted using Qiagen® RNeasy Mini kit according to manufacturer's instruction. Real-time RT-PCR reactions were performed at National Animal Health Diagnostic Center, Sebeta using an Applied Biosystems 7500 Fast Real-Time PCR thermo cycler. A primer probe combination from conserved region of M gene APMV-1 F M+4100 5'-AGT GAT GTG CTC GGA CCT TC-3', APMV-1 R M-4220 5'-CCT GAG GAG AGG CAT TTG CTA-3' and Probe APMV-1 Probe M+4169 5'- FAM- TTC TCT AGC AGT GGG ACA GCC TGC BHQ-3' (Wise et al. 2004) was used to amplify and detect all NDV isolates. The M-gene positive samples were re-run by a second primer-probe set targeting the F gene Forward F+4829 5'-GGT GAG TCT ATC CGG ARG ATA CAA G-3', Reverse Primer F- 4939 5'-AGC TGT TGC AAC CCC AAG-3' and Probe F+4894 5'-FAM-AAG CGT TTC TGT CTC CTT CCT CCA-BHQ-3' (Wise et al. 2004) to specifically amplify and detect only pathogenic strains of NDV. Twenty nine F gene Positive allantoic fluids were submitted to Instito Zooprofilattico Sperimentale delle Venezie (OIE/FAO Reference Laboratory for AI and ND), Padua,Italy for further analyses.

## 2.5 Nucleotide Sequencing and phylogenetic analysis

Total RNA was extracted from infective allantoic fluids using Qiagen® RNeasy Mini kit according to manufacturer's instruction. Amplification was performed with primers NOH-For (5'-TAC ACC TCA TCC CAG ACA GG-3') and NOH- Rev (5'-AGT CGG AGG ATG TGT TGG CAG C-3') which encompass 260 bp region of the fusion (F) gene. Amplicons were then purified with ExoSAP-IT (USB Corporation, Cleveland, OH). Sequencing was carried out by using Big Dye Terminator v 3.1 cycle sequencing kit (Applied Biosystems, Foster City, CA, USA) in a 16-cappilary ABI PRISM 3131x Genetic Analyzer (Applied Biosystems, Forster City, CA,USA).

For complete sequencing of F gene, RNA was reverse transcribed with the Superscript III Reverse Transcriptase kit (Invitrogen, Carsbad, CA, USA) and amplified using primer pairs described in Table 1. Sequence data were assembled and edited with Seq Scape software v2.5 (Applied Biosystems). To determine the phylogenetic relationships, the sequences of the F gene were compared to the corresponding region of representative viruses of class II available in GenBank. Alignment and comparison of the nucleotide and amino acid sequences were performed using Clustal W in MEGA 5.0. Maximum likelihood (ML) trees were estimated using best-fit general time-reversible(GTR) model of nucleotide substitution with gamma-distributed rate variation among sites, and a heuristic SPR branch-swapping search available in PhyML version 3.0 (Guindon and Gascuel 2003). A bootstrap resampling process (100 replications) was employed to assess the robustness of individual nodes of phylogeny.

**Table 1** Primers set for amplification of complete F-gene

| Primers | Sequence 5′ to 3′ |
|---|---|
| MF3710-F | TGA AAA CGA CGG CCA GTC AAA GCT GTA DGG TTG TG |
| MF4650-R | CAG GAA ACA GTA TGA CCA AGA GGC CTG CCR TCA A |
| F5100-F | TGA AAA CGA CGG CCA GTA TGC AGC ART TTG TYA AT |
| HF011-R | CAG GAA ACA GTA TGA CCT ARG TAA TRA GAG CRG ATG |
| HF005-F | TGA AAA CGA CGG CCA GTA GAC YGA AGG CGC ACT YAC |
| HH008-R | CAG GAA ACA GTA TGA CCA GRG CCA CYT GCT TRT ATA |
| F-4514-F | TGA AAA CGA CGG CCA GTG TAG AAG ADT YTG GAT CC |
| F-5218-R | CAG GAA ACA GTA TGA CCG AAT ACY GTA GTC AAY TCR G |
| HF009-R | CAG GAA ACA GTA TGA CCA GGT GGC ACG CAT ATT ATT |
| F-5757-F | TGA AAA CGA CGG CCA GTA GAT RAC AAC ATG TAG RTG |
| F-6449-R | CAG GAA ACA GTA TGA CCG GCT AAC YGC RCG GTC CAT |

# 3. RESULTS

## 3.1 Marketing practices

Chickens present at each market were estimated to be 300–500 on the days of visit. They are originated from different districts across the country, most of which were collected by traders from big towns such as: Sodo, Dessie, Shashemane, Jimma, Ambo, Hosaina and transported on public transport (bus) to Addis Ababa. The majority (99%) of traders mentioned that there is no habit of cleaning buses before loading and after unloading chickens owing to the fact that no vehicles is designed for poultry transportation alone. After arrival at the market sites, chickens are sold to wholesalers in the biggest live bird market site (Merkato) where birds from different origins are placed in the same cages (mixed) and sold either to traders in other live poultry markets in the city or to the consumers on spot. Ninety nine percent of the traders clean the holding cages once a week with clean water and the remaining 1% clean every day. All the traders responded that they dispose manure and dead birds either within the markets or on roadside. None of them seek for veterinary services when chickens are sick.

## 3.2 Laboratory results

Out of the 146 pooled samples tested, 44 were positive for NDV by both HA/HI and M gene real-time RT-PCR tests. Out of these, 29 isolates were identified as virulent based on the real-time RT-PCR of F gene. Among these, 14 Tracheal swab pools (TS) were positive though the corresponding Cloacal swab pools (CS) were negative. Similarly, 3 CS were positive though the corresponding TS were negative. However, the remaining 12 swab pools have shown direct correlation (12 TS and the corresponding 12 CS were positive) and were not considered as separate entity to avoid double counting of the samples (Table 2). The identification, pathotyping and genotyping of the 29 isolates were confirmed by genetic sequencing of the 260 bp region of fusion (F) gene and the complete F gene sequence of two isolates (KJ958913_13VIR3936-1 and KJ958914_13VIR3936-27) collected from the market sites Saris and Kotebe. Since all the viruses showed a high similarity at the 260 bp region of the F gene (Table 2), the selection was based on the sequencing quality (i.e. absence of degenerate bases).

The isolates were identified as virulent as deduced amino acid sequence of the fusion protein cleavage site indicated the presence of motif $^{112}$RRQKR$^{116}$*F$^{117}$ typical of NDV virulent strains. Phylogenetic analysis of the 260 fragment of the F gene of all the 29 sequenced isolates (Additional file 1: Figure S1) and complete coding region of the F gene of two representative viruses (Figure 2) indicated that the viruses are clustered together (similarity ranged from 98% to 100%) within the new sub-genotype VIf.

**Table 2** Origin of positive swab pools and their genetic identity with reference sequence

| Market sites from where the samples were collected | Number of M-gene real time RT PCR positive swab pools | Number of F-gene(virulent) real time RT PCR positive swab pools | | | | Genetic identity with the reference sequence at the 260 bp region of the F gene |
|---|---|---|---|---|---|---|
| | | Total | Both TS &CS | Only TS | Only CS | |
| Shola | 12 | 8 | 4 | 4 | 0 | 98 to 100% |
| Meri | 9 | 5 | 4 | 1 | 0 | 98.7 to 100% |
| Saris | 7 | 3 | 2 | 0 | 1 | 99.3 to 100% |
| Merkato | 10 | 9 | 2 | 5 | 2 | 100% |
| Kotebe | 6 | 4 | 0 | 4 | 0 | 100% |

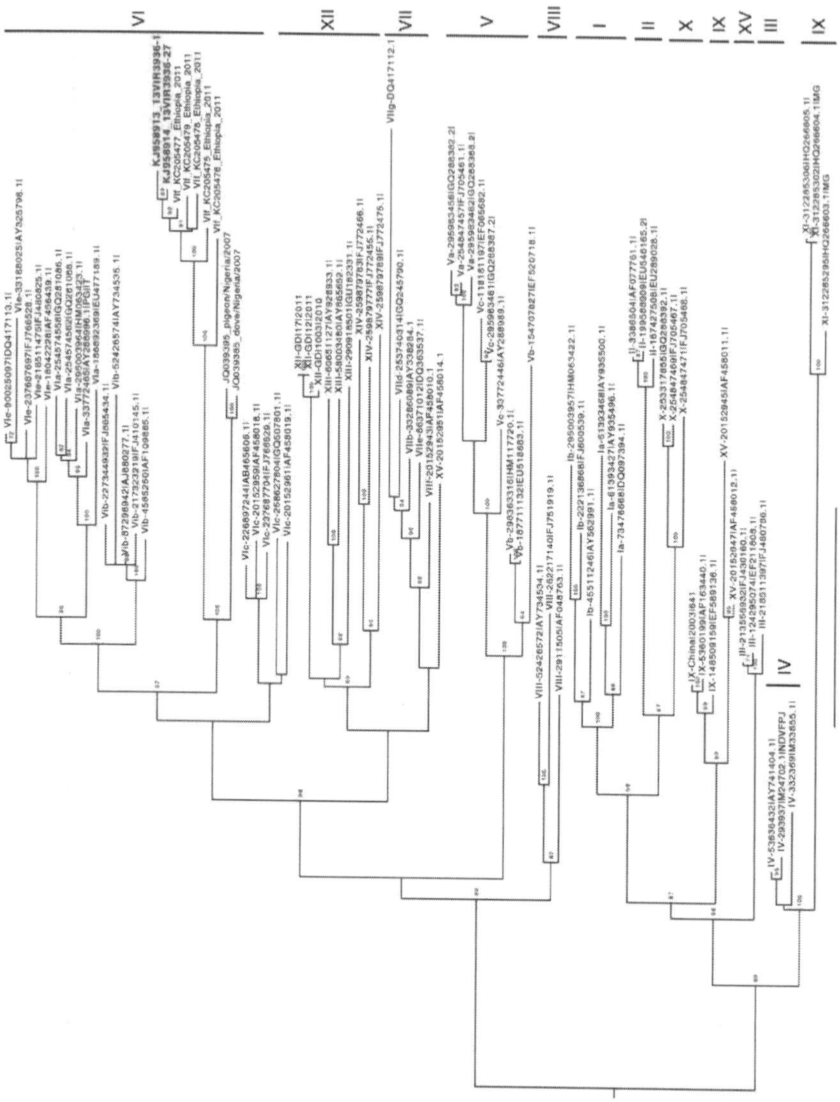

**Figure 2** Phylogenetic tree of the complete nucleotide sequence of the F gene for two representative isolates. ML Phylogenetic tree of complete nucleotide sequence of Class II Newcastle diseases viruses. Ethiopian isolates included in this analysis (KJ958913_13VIR3936-1 and KJ958914_13VIR3936-27) are labeled in red. The Nomenclature system used in the phylogenetic tree is based on (Diel et al.2012). The numbers at branch points represent bootstrap values.

# 4. DISCUSSION

In this study, we assessed biosecurity practices in market sites in Addis Ababa that contribute to the spread of the diseases along market chain. We observed several practices that could promote Newcastle disease and other infectious diseases transmission among birds. These included limited cleaning and disinfection of the market places, holding cages, mixing of birds from different origins, inappropriate disposal of sick and dead birds and poor waste disposal mechanisms. Earlier reports indicated that marketing systems play a considerable role in the dissemination of diseases over wide geographical areas in a relatively short period of time in Ethiopia (Gebreab 1995). Similar observation was reported from Addis Ababa live poultry market sites (Shewantasew et al. 2012). The current finding is also in agreement with a report from Nairobi live bird markets where poor biosecurity practices were suggested to play major roles for transmission of avian influenza viruses (Munyua et al. 2012). These factors were found to be associated with transmission of low pathogenic avain influenza viruses in markets in North America (Garber et al. 2007).

The current study showed that, virulent viruses with multiple basic amino acids at cleavage site of F protein $^{112}$RRQKR$^{116}$*F$^{117}$ are circulating in village chickens. The finding is in agreement with earlier studies which revealed that, the majority of the virus strains circulating in the village chickens in Ethiopia are virulent strains (Chaka et al. 2012). Virulent Newcastle disease virus (vNDV) is endemic in many countries of Africa, North, Central, and South America, and outbreaks of ND are frequently reported to the World Organization of Animal health (Office of International Epizootes 2011).

According to Diel et al. (2012) classification, our phylogenetic analysis of the 260 fragment of the fusion gene of all the 29 sequenced isolates (Additional file 1: Figure S1) indicated that the viruses are grouped in the new sub-genotype VIf class II viruses, and grouped with NDV identified in Ethiopia (de Almeida et al. 2013; Chaka et al. 2013). The genotype was confirmed by analysis of complete coding region of the F gene of two representative viruses (Figure 2), which showed 95% to 99.3% similarity with the NDV isolated in 2011 suggesting the persistence of this sub-genotype in poultry in the country. While the 260 nucleotide sequences of samples collected from the market sites of Merkato and Kotebe were all identical, sequences of viruses from Saris, Shola and Meri showed a nucleotide identity ranging from 99.3% to 100%, from 98% to 100% and from 98.7% to 100%, respectively). Which revealed that, identical isolates are circulating in livebird markets in Addis Ababa. This could be associated with the poor biosecurity practices and the use of similar marketing channel in these market sites. However, Fentie et al. (2013) identified genotype VII viruses in the North western Ethiopia, indicating the co-circulation of different genotypes in Ethiopia. Genotype VI, include viruses that have been isolated from multiple avian species (Alexander 2011; Czeglédi et al. 2006; Kim et al.2008). Viruses of this group are important because of their frequent association with doves and pigeons and consequent risk for introduction in to poultry flocks (Alexander 2011; Kim et al. 2008). It is interesting to note that Ethiopian viruses analyzed here grouped with viruses collected in Nigeria from pigeon and doves (identity of about 90%), suggesting that this genotype may have emerged in Africa as a result of transmission from pigeon to poultry.

In conclusion, the current study revealed that village chicken flocks are endemically infected with virulent Newcastle disease virus, which could pose a threat to commercial poultry farms. This study highlights the importance of implementing surveillance and molecular investigations of Newcastle disease, in village chickens and wild birds, along with improved biosecurity measures in the live poultry markets in Ethiopia.

## ACKNOWLEDGEMENTS

This study was supported by National Animal Health Diagnostic center, Ethiopia. The authors would like to thank the molecular Biology Laboratory staff of NAHDIC, Ethiopia, and Istituto Zooprofilattico Sperimentale delle Venezie,Italy for their technical assistance and support in processing of the samples and genetic analysis of the isolates.

## COMPETING INTERESTS

The authors have declared that no competing interests exist.

## AUTHORS' CONTRIBUTIONS

DD conceived and designed the study, participated in data collection, laboratory analysis and drafted the manuscript, MK participated in data collection, laboratory analysis and helped in drafting of the manuscript, RB and MS participated in data collection and laboratory analysis , AF carried out the sequencing, sequence alignment and phylogenetic reconstruction and revised the manuscript, AH participated in sequence alignment and phylogenetic reconstruction, TR participated in design and coordination of the work. HC participated in drafting and edition of the manuscript. All authors read and approved the final manuscript.

## REFERENCES

1 Alexander DJ: Newcastle disease in the European Union 2000 to 2009. *Avian Pathol* 2011, 40(6):547-558. 10.1080/ 03079457. 2011. 618823

2 Alexander DJ, Senne DA: Newcastle disease, other avian paramyxoviruses, and pneumovirus infections. In *Diseases of Poultry*. 12th edition. Edited by: Saif YM, Fadly A, Glisson JR, Mc D, Larry R, Nolan LK, Swayne D. Ames, IA: Blackwell Publishing; 2008:750-798.

3 Central Statistical Agency: *Agricultural Sample Survey (2011/12) Volume II, Report on Livestock and Livestock characteristics, Statistical Bulletin, 505*. Addis Ababa, Ethiopia; 2012.

4 Chaka H, Goutard F, Bisschop SPR, Thompson PN: Seroprevalence of Newcastle disease and other infectious diseases in backyard chickens at markets in Eastern Shewa zone, Ethiopia. *Poult Sci* 2012, 91: 862-869. 10.3382/ps.2011-01906

5 Chaka H, Goutard F, Gil P, Abolnik C, Servan De Almeida R, Bisschop S, Thompson PN: Serological and molecular investigation of Newcastle disease in household chicken flocks and associated markets in Eastern Shewa zone, Ethiopia. *Tropl Anim Health Prod* 2013, 45(3):705-714. 10.1007/s11250-012-0278-y

6 Czeglédi A, Ujvári D, Somogyi E, Wehmann E, Werner O, Lomniczi B: Third genome size category of avian paramyxovirus serotype 1 (Newcastle disease virus) and evolutionary implications. *Virus Res* 2006, 120(1–2):36-48.

7 De Almeida RS, Hammoumi S, Gil P, Briand FX, Molia S, Gaidet N, Cappelle J, Chevalier V, Balanc G, Traore A, Grillet C, Maminiaina OF, Guendouz S, Dakouo M, Samake K, Bezeid OEM, Diarra A, Chaka H, Goutard F, Thompson P, Martinez D, Jestin V, Albina E: New avian paramyxoviruses type I strains identified in Africa provide new outcomes for phylogeny reconstruction and genotype classification. *PLoS One* 2013, 8(10):76413. 10.1371/journal.pone.0076413

8 Dessei T, Ogle B: Village poultry production system in the central highlands of Ethiopia. *Tropl Anim Health Prod* 2001, 33: 521-537. 10.1023/A:1012740832558

9 Diel DG, Da SLHA, LiuH WZ, Miller PJ, Afonso CL: Genetic diversity of avian paramyxovirus type 1: Proposal for a unified nomenclature and classification system of Newcastle disease virus genotypes Infection. *Genet Evol* 2012, 12: 1770-1779. 10.1016/j.meegid.2012.07.012

10 Fentie T, HeidariA AR, Kassa T, Capua I, Cattoli G, Sahle M: Molecular characterization of Newcastle disease viruses isolated from rural chicken in northwest Ethiopia reveals the circulation of three distinct genotypes in the country. *Tropl Anim Health Prod* 2013, 10: 11250-11263.

11 Garber L, Voelker L, Hill G: Description of live poultry markets in the United States and factors associated with repeated presence of H5 /H7 low-pathogenicity avian influenza virus. *Avian Dis* 2007, 51: 417-420. 10.1637/7571-033106R.1

12 Gebreab F: *Zoonotic diseases in Ethiopia. Ethiopian Society of Animal Production*. Addis Ababa, Ethiopia; 1995.

13 Glickman RL, Syddall RJ, Iorio RM, Sheehan JP, Bratt MA: Quantitative basic residue requirements in the cleavage-activation site of the fusion glycoprotein as a determinant of virulence for Newcastle disease virus. *J Virol* 1988, 62: 354-356.

14 Guindon S, Gascuel O: A simple, fast, and accurate algorithm to estimate large phylogenies by maximum likelihood. *Syst Biol* 2003, 52: 696-704. 10.1080/10635150390235520

15 Kattenbelt JA, Stevens MP, Gould AR: Sequence variation in the Newcastle disease virus genome. *Virus Res* 2006, 116(1–2):168-184.

16 Kim LM, King DJ, Guzman H, Tesh RB, Travassos DA, Rossa APA, Bueno R, Dennet JA, Afonso CL: Biological and phylogenetic characterization of pigeon paramyxovirus serotype 1 circulating in wild North American pigeons and doves. *J Clin Microbiol* 2008, 46(10):3303-3310. 10.1128/JCM.00644-08

17 Mayo: A summary of taxonomic changes recently approved by ICTV. *Arch Virol* 2002, 147: 1655-1656. 10.1007/s007050200039

18 Mazengia H: Review on major viral diseases of chickens reported in Ethiopia. *J Infect Dis Immun* 2012, 4(1):1-9.

19 Munyua MP, Githinji WJ, Waiboci WL, Njagi ML, Arunga G, Mwasi L, Mbabu MR, Macharia MJ, Breiman FR, Njenga KM, Katz AM: Detection of influenza A virus in live bird markets in Kenya, 2009–2011. *Influenza Other Respi Viruses* 2012, 10: 1750-2659.

20 Nasser M: *Oral Newcastle disease vaccination trial and studies of Newcastle disease in Ethiopia. (M.Sc. Thesis)*. AddisAbaba University and Freie Universitat, Addis Ababa, Ethiopia and Berlin, Germany; 1998.

21 Office of International Epizootes: *World Animal Health Information Database (WAHID) Interface*. 2011. http://web.oie.int/wahis/public.php?page_home&WAHIDPHPSESSID4d95e8777ff8b8a615b31617ac4c7835

22 Office of International Epizootes: *Newcastle Disease*. Manual of Diagnostic Tests and Vaccines for Terrestrial Animals, Paris, France; 2012. chapter 2.3.14

23 Shewantasew M, Melesse B, MesfinS MB: Assessment of Bio-security situation and practices in live poultry Markets of Addis Ababa, Ethiopia. *Asian J Anim Vet Adv* 2012, 7(5):421-433.

24 Spradbrow PB: Epidemiology of Newcastle Disease and the Economic of its control. In *Proceedings of international workshop on Poverty eradication and promotion of gender equality, held at Livestock Research Institute 26–26 March1999*. Denmark: Tune landboskole; 1999:165-173.

25 Steward M, Vipond IB, Millar NS, Emmersion PT: RNA editing in Newcastle disease virus. *Journal of General Virology* 1993, 74: 2539-2547. 10.1099/0022-1317-74-12-2539

26 Tadelle , Jobre Y: A review of the importance and control of Newcastle disease in Ethiopia. *Ethiop Vet J* 2004, 1: 71-81.

27 Toyoda T, Sakaguchi T, Imai K, Inocencio NM, Gotoh B, Hamaguchi , Nagai Y: Structural comparison of the cleavage-activation site of the fusion glycoprotein between virulent and avirulent strains of Newcastle disease virus. *Virology* 1987, 158: 242-247. 10.1016/0042-6822(87)90261-3

28 Wise MG, Suarez DL, Seal BS, Pedersen JC, Senne DA, King DJ, Kapczynski DR, Spackman E: Development of a real-time reverse-transcription PCR for detection of Newcastle disease virus RNA in clinical samples. *J Clin Microbiol* 2004, 42(1):329-338. 10.1128/JCM.42.1.329-338.2004

# CHAPTER 5

# An Alternative Vaccination Approach for The Prevention of Highly Pathogenic Avian Influenza Subtype H5N1 in The Red River Delta, Vietnam—A Geospatial-Based Cost-Effectiveness Analysis

**Chinh C. Tran [1,2,*], John F. Yanagida [1], Sumeet Saksena [2] and Jefferson Fox [2]**

[1] Department of Natural Resources and Environmental Management, University of Hawaii at Manoa, 1910 East-West Road, Honolulu, HI 96822, USA

[2] East-West Center, 1601 East-West Road, Honolulu, HI 96848, USA

## ABSTRACT

This study addresses the tradeoff between Vietnam's national poultry vaccination program, which implemented an annual two-round HPAI H5N1 vaccination program for the entire geographical area of the Red

River Delta during the period from 2005–2010, and an alternative vaccination program which would involve vaccination for every production cycle at the recommended poultry age in high risk areas within the Delta. The *ex ante* analysis framework was applied to identify the location of areas with high probability of HPAI H5N1 occurrence for the alternative vaccination program by using boosted regression trees (BRT) models, followed by weighted overlay operations. Cost-effectiveness of the vaccination programs was then estimated to measure the tradeoff between the past national poultry vaccination program and the alternative vaccination program. *Ex ante* analysis showed that the focus areas for the alternative vaccination program included 1137 communes, corresponding to 50.6% of total communes in the Delta, and located primarily in the coastal areas to the east and south of Hanoi. The cost-effectiveness analysis suggested that the alternative vaccination program would have been more successful in reducing the rate of disease occurrence and the total cost of vaccinations, as compared to the national poultry vaccination program.

## KEYWORDS

HPAI H5N1; alternative vaccination program; national poultry vaccination program; Red River Delta

## 1. INTRODUCTION

The Highly Pathogenic Avian Influenza (HPAI) subtype H5N1 has had serious, detrimental effects on the economy and human health in Vietnam since the first reported outbreak on 27 December 2003 [1]. Millions of poultry were culled due to disease occurrences, causing an estimated economic loss of 3 trillion VND (approximately U.S. $187.15 million) [2,3]. The average growth rate of poultry population was reduced from 7.6% for the period 2000–2003 (before HPAI H5N1 occurrence) to 3.8% for the period 2003–2006 (during HPAI H5N1 disease occurrence) [4]. Market demand and price decreases caused further economic losses to poultry producers [5,6]. The disease also seriously affected human health. By 19 November 2010, a total of 119 human cases of HPAI H5N1 were reported, with 59 deaths [7].

Financial support from many international organizations, including the Food and Agriculture Organization (FAO) of the United Nations, the World Bank, the World Health Organization, and others, helped to contain the disease as designed in the Integrated National Operational Program for Avian and Human Influenza (OPI) which is also known as the *Green Book*[8]. The overall objective of the OPI was to reduce the health risk to humans by controlling the HPAI H5N1 disease at the source in domestic poultry. The national poultry vaccination campaign was a key strategic plan in the OPI [8] (we hereby use the term "national poultry vaccination program" throughout the study). After the pilot vaccinations implemented in Nam Dinh and Tien Giang provinces in August 2005, a mass vaccination campaign was conducted nationally from late September to the beginning of November, 2005 and continued until the end of 2010. The vaccination was applied nationally in two rounds per year with the first round from April–May and the second round from October–November. The vaccination program continued in later years, but on a smaller scale, and was primarily determined and implemented by provincial authorities.

Vaccination has been shown to be a viable means of protection against the HPAI H5N1 virus [9]. While the campaign was carried out only twice a year in April–May and October–November, poultry production, however, occurs all year round. A previous study confirmed that November to January and April to June were the periods that are most vulnerable for disease occurrence [5]. A sizeable proportion of poultry population remained unvaccinated at different times of the year. Therefore, unvaccinated poultry between the two rounds of vaccinations will be at risk of infection. As a result, the disease has been repeatedly reported over the years. Circulation of the HPAI H5N1 virus was found mostly in unvaccinated waterfowl, especially ducks [10,11]. Additionally, the cost of the vaccination program was estimated to be approximately U.S. $10 million per round [7,12,13].

Therefore, it is critical to explore implications of an alternative vaccination program, which is likely to be more successful in containing and preventing the disease from recurrence in the Red River Delta, Vietnam, and reducing total vaccination costs as compared to the national poultry vaccination program implemented in the period 2005–2010. This study focuses on the Red River Delta of Vietnam. This area has been identified as a high-risk area for the disease [10,14]. Vaccination was implemented across all geographical areas for all provinces in the Delta under the national poultry vaccination program.

A number of studies have identified factors affecting the occurrence and spread of the HPAI H5N1 virus in Vietnam, in general, and in the Red River Delta, in particular. It was suggested that higher average monthly temperatures and poultry density in combination with lower average monthly precipitation, humidity, and elevation significantly affected the occurrence of HPAI H5N1 in the Delta [5]. Other factors linked with the disease at the national level were a higher proportion of land used for rice paddy fields and aquaculture, increases in production, trade and movement of live poultry, and the expansion of free-grazing ducks [14,15,16,17].

Given these findings, it is not likely that all areas within the Red River Delta are equally susceptible to the disease. We propose an alternative vaccination program which involves shifting from a less frequent blanket vaccination for the entire Delta to more frequent vaccination in high probability areas for disease occurrence within the Delta. This modification would involve vaccination for every production cycle at the recommended poultry age in high risk areas. Two key questions emerge with this proposal: (i) Where are the higher probability areas (focus areas) for the alternative vaccination program? (ii) Is it beneficial for the Government to switch to the alternative vaccination program in terms of the cost-effectiveness of vaccination programs? To answer questions and fulfill the objective, this study (i) identifies the focus areas for the alternative vaccination program to be implemented in the Red River Delta, and (ii) estimates the tradeoff between the national poultry vaccination program and the alternative vaccination program based on the cost-effectiveness analysis of vaccination programs.

# 2. METHODS

## 2.1. Study Area and Data Sources

This study focuses on the Red River Delta of Vietnam (Figure 1), which represents one of the two largest flood plains in Vietnam. The Delta includes two large river systems—the Red river and Thai Binh river systems—that support agricultural and livestock activities. The Red River Delta includes eight provinces and two municipalities, the capital city of Hanoi and the main port of Hai Phong. The Delta plays an important role and interacts with a wide range of environmental and socioeconomic sectors, including industry, commerce, services, agriculture, tourism, *etc.* Livestock production is among the main activities in the Delta, including poultry, pig, and cow husbandry. Poultry production has faced serious problems caused by the HPAI H5N1 disease.

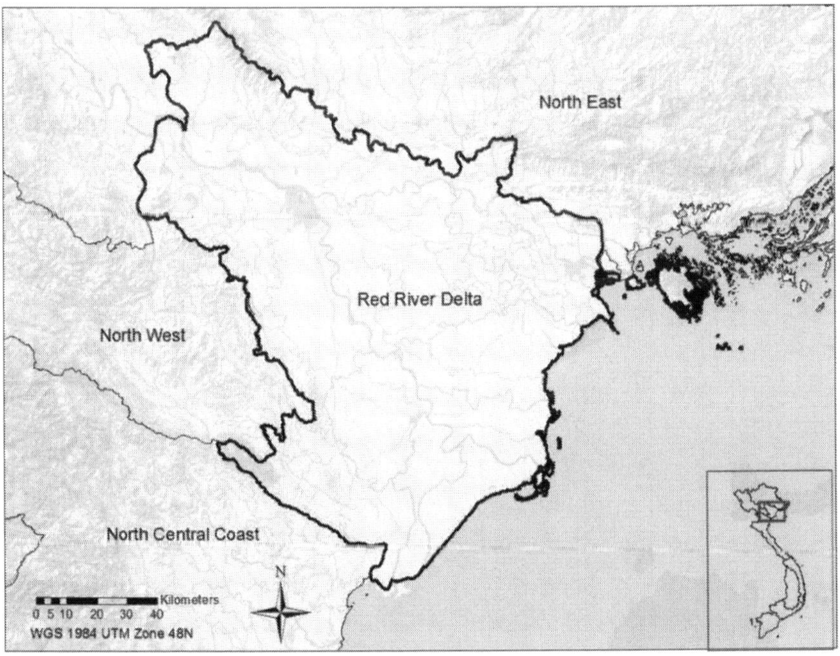


**Figure 1.** The study area—the Red River Delta, Vietnam.

The HPAI H5N1 outbreak data are routinely collected and reported by the Vietnam Department of Animal Health. The Red River Delta has been identified as a high-risk area for the disease [10,14]. The Delta has been severely affected by three out of the five large epidemic waves of HPAI H5N1 outbreaks reported in Vietnam. These three waves included the first wave, which occurred from December 2003–February 2004, the third wave was recorded from October–December 2005, and the fifth wave reported from May–September 2007 [14,18]. A number of outbreaks in the second epidemic wave which was from November 2004 to March 2005 were also reported in the Red River Delta despite the primary effects occurring in the Mekong Delta. Other sporadic outbreaks occurred over the period from December 2003 to the present, with recent reported outbreaks in Bac Ninh and Nam Dinh provinces in January and February 2014 [1]. Although the disease occurred in the Delta with the first epidemic wave reported from the end of 2003, the dates and locations of occurrences were not formally reported until the end of March 2004 [14]. Therefore, this study analyzed reported disease data for the period starting from the end of March, 2004, to the end of December, 2007, which included four large epidemic waves of HPAI H5N1 outbreaks with 267 confirmed HPAI H5N1 outbreaks in the Red River Delta. The data were reported at the commune level and coded as 1 if the disease was found or 0 if there was no disease reported.

Other risk factors used for the analysis were identified based on earlier studies of HPAI H5N1 in Vietnam. These included the percentage of land used for rice paddy fields and aquaculture [14,19], chicken and water bird density [4,14,15] and elevation [5,14,17]. Land use diversity was found to be significantly associated with disease occurrence [19]. Land use/land cover is dominated by forests and permanent vegetation in high elevation areas, and characterized by agriculture and a mixed uses of land in low elevation areas [17]. Disease occurrence was also found around heavily-populated cities in different regions in Vietnam such as Ho Chi Minh city in the South, Da Nang city in the central part, and Hanoi and Hai Phong cities in the north [14]. Therefore, two other land use factors, characterizing built-up and forest/perennial trees features, were also included.

These variables, percentage of land use for rice paddy fields, aquaculture, built-up and forest/perennial trees, and chicken and water bird density, were measured at the commune level and obtained from The 2006 Vietnam Rural, Agricultural and Fishery Census provided by the East West Center–National Science Foundation project, "Coupled Natural-Human Systems and Emerging Infectious Diseases: Anthropogenic environmental change and avian influenza in Vietnam". Elevation data were obtained from the Shuttle Radar Topography Mission (SRTM) 90-m resolution Digital Elevation Model (DEM) [20]. These data were then retrieved for each commune and merged with other data using commune codes for the statistical analysis. Remotely sensed Landsat TM/ETM+ Bands 1–5, 7 data that cover the study area were downloaded from the USGS EROS Data Center [21]. The Red River Delta is covered by four Landsat tiles: P126R045, P126R046, P127R045, and P127R046 (Figure 2).

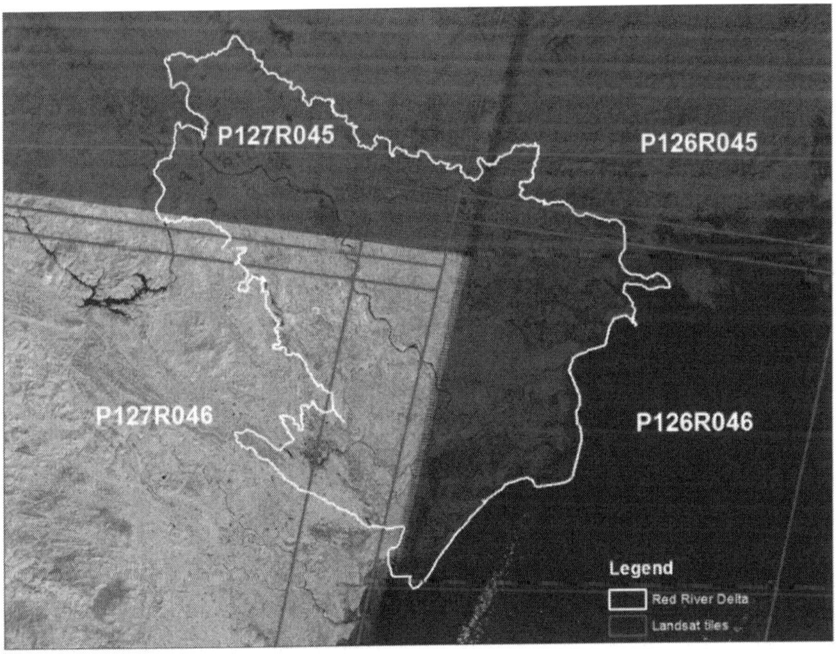


**Figure 2.** Landsat tiles covering the Red River Delta of Vietnam.

## 2.2. Statistical Analysis

For the first objective involving identification of focus areas for the alternative vaccination program in the Red River Delta, this study adopted the *ex ante* analysis framework. Outbreak data were divided into two datasets. The first dataset contained the data before the launch of the national poultry vaccination program in 2005, which included the second and the third epidemic waves, with a total of 193 outbreaks. The second dataset consisted of outbreak data that occurred after 2005 which comprised the fifth epidemic wave with 74 outbreaks. The use of the first dataset was to identify focus locations for the alternative vaccination program. The second dataset would be used to evaluate the cost and effectiveness of the vaccination program in analyzing the tradeoff between the national poultry vaccination program and alternative vaccination program.

### 2.2.1. *Ex Ante* Analysis

For *ex ante* analysis, weighted overlay analysis was applied to identify the focus locations for the alternative vaccination program. This method has been considered as one of the most suitable techniques and is frequently used for site selection and suitability models in spatial analysis [22]. It has been widely applied in several fields, e.g., disease management, climate change, habitat conservation, sustainable ecosystems, and land-use planning, *etc.* [23,24,25,26,27]. This technique requires that all input factors are classified into different groups and weighted to determine their weight accordingly. The analytical procedure for weighted overlay analysis in this study involved a two-stage process: (1) boosted regression trees (BRT), followed by (2) weighted overlay operations.

The first step is to run BRT to determine the appropriate weighting for the main factors that were previously found to have significant effects on disease occurrence, including water bird density, chicken density, elevation, and land use/land cover (Table 1). BRT utilizes a combination of decision trees and boosting algorithms to improve prediction accuracy through an iterative process [28,29]. This method has been popularly applied for predicting the distribution of HPAI

H5N1 disease [19,30,31,32,33]. Several combinations of the learning rate and tree complexity were tested to choose the best setting for evaluating model performance which was determined through cross-validation (CV). The final model setting was the combination of a tree complexity of 4 and a learning rate of 0.005, with a bag fraction of 0.75, which were also previously used in [30]. The results present which factors appear to have more influence on the disease occurrence in the Red River Delta.

**Table 1.** Critical factors affecting the disease occurrence.

| Factors | Data Sources | Attribute Values of Factors |
|---|---|---|
| Water bird density (heads/km$^2$) | The 2006 Vietnam Rural, Agricultural and Fishery Census | 0–892<br>893–2097<br>2098–4299<br>>4299 |
| Chicken density (heads/km$^2$) | The 2006 Vietnam Rural, Agricultural and Fishery Census | 0–1738<br>1739–3992<br>3993–9472<br>>9472 |
| Elevation (m) | SRTM 90-m resolution DEM: http://srtm.csi.cgiar.org/ [20] | ≤5<br>>5–15<br>>15–200<br>>200 |
| Land use/land cover (%) | The 2006 Vietnam Rural, Agricultural and Fishery Census | Agriculture<br>Aquaculture<br>Built-up<br>Forest/perennial trees |

Each factor has its own characteristics that also have impacts on the disease. The next step is to measure the weights of attribute values of each factor. The land use/land cover included land for agriculture, aquaculture, built-up purposes, and forest/perennial tree areas. The Red River Delta topography was reclassified into four groups of elevation (above 200 m, from 15 m to 200 m, from 5 m to 15 m, and less than 5 m) and coded from 1 to 4, respectively, to represent upland, midland, lowland, and coastal areas (see [5]). Other factors were categorized into four groups by using the commonly-used Jenks natural breaks classification method [34] in ArcGIS to group data into categories (Table 1).

The second step determined the potential focus area for the vaccination program by performing weighted overlay analysis in

ArcGIS 10.1 (ESRI, Redlands, CA, USA). The analysis was operated in raster layers. Therefore, vector layers detailing water bird density and chicken density at the commune level were converted to corresponding raster layers. The elevation data were originally stored in raster format. For the spatial distribution of land use/land cover, a support vector machine (SVM) was applied to classify remotely-sensed imagery data in the Red River Delta into four categories that represent built-up, agriculture, forest/perennial trees, and water areas. The SVM method has been successfully applied in several studies on biophysical tasks, land cover/land use including vegetation, agriculture, and impervious surfaces, such as urban areas, *etc.* [35,36,37]. The classification process was performed using ENVI version 4.8 (Exelis Visual Information Solutions, Boulder, CO, USA) and ArcGIS version 10.1 (ESRI, Redlands, CA, USA). Outputs were the classification maps of land use/land cover for each subset and were mosaiced together to produce the final land use/land cover classification map for the Red River Delta.

All raster layers were then reclassified using the corresponding weight obtained from BRT results for the weighted overlay operations processed in ArcGIS version 10.1 to find the suitable areas for the alternative vaccination program. The analysis provides suitability maps with suitability scores in integer numbers scaled from 0 to 100. The higher suitability scores represent the higher probability of contracting the HPAI H5N1 disease. Areas with higher suitability scores are suggested as good candidates where the alternative vaccination program should be focused.

### 2.2.2. The Cost-Effectiveness Analysis

For the tradeoff between the past national poultry vaccination program and the alternative vaccination program, the cost and effectiveness of the vaccination program were estimated for each program. The second dataset which included data for the fifth epidemic wave with 74 affected communes was used in addition to the first dataset. To implement the vaccination program, the government is responsible for all the costs, including the costs of vaccine, labor, and other costs associated with vaccination. The costs for the vaccination program are the product of the number of birds vaccinated, the cost of vaccination per bird, and the number of vaccination rounds per year. When the disease occurs, the Vietnamese government implements the stamping out method, which culls all birds in affected communes and emergency vaccination is deployed to vaccinate all birds in surrounding communes in order to contain and prevent the disease from spreading. These are extra costs of the vaccination program. The total cost of the vaccination program is comprised of the cost of the government vaccination program, the cost of emergency vaccination, government compensation for birds culled, and farmer's losses because of the value difference between market price and the government's compensation when the disease occurs.

$$\text{Cost} = A \times C \times N + B \times C + I \times G + (P - G) \times I \quad (1)$$

where A is the number of birds vaccinated; C is the costs of vaccination per bird; N is the number of vaccination rounds per year; B is the number of birds vaccinated because of emergency vaccination; I is the number of birds culled due to the disease occurrence; G is the government compensation per bird culled; and P is the market price per bird.

The effectiveness of the vaccination program is the measure of proportionate reduction in the rate of disease occurrence as the result of the vaccination program. This can be achieved through the calculation based on the relative risk of the disease [38].

$$\text{Effectiveness} = \frac{\text{ARU-ARV}}{\text{ARU}} \times 100 \quad (2)$$

where ARU and ARV are, respectively, the infection rates before and after the launch of the vaccination program. The infection rate is the number of affected communes divided by the total communes in the Red River Delta.

# 3. RESULTS AND DISCUSSION

## 3.1. Results

### 3.1.1. *Ex Ante* Analysis

The BRT results (see Table 2) suggest that water bird density had the largest effect on disease occurrence with the weight estimated at 19%. Ducks, as a reservoir host for the HPAI H5N1 virus, have been discussed in earlier studies [10,16,17,39,40,41,42]. The number of recorded duck-related disease occurrences steadily increased from 11% in 2003/2004 to its peak of 78% in 2006/2007 [10]. The next highest weight factors were land used for rice paddy field, elevation, land used for aquaculture, land used for built-up purposes, chicken density, and land used for forest/perennial trees. Land used for forest/perennial trees had the smallest effect with 2% weight.

**Table 2.** Estimated weight of each factor to the HPAI H5N1 occurrences.

| Variable | Weight (%) |
|---|---|
| Water bird density | 19 |
| Land used for rice paddy field | 18 |
| Elevation | 18 |
| Land used for aquaculture | 17 |
| Land used for built-up purposes | 14 |
| Chicken density | 12 |
| Land used for forest/perennial trees | 2 |

Table 3 reveals the weight of attribute values of each factor; lower water bird density and chicken density were found to have higher weight as compared to other groups within each factor. Specifically, water bird density ranging from 893–2097 heads/km$^2$ had the highest weight of 76% for the water bird density factor. Chicken density attributes ranging from 0–1738 heads/km$^2$ showed the highest weight of 37%. Elevation less than 5 m was estimated to have a 73% weight. These were flat plain areas where rice production was the predominant agricultural activity in the Red River Delta. Lowlands with elevation ranging from 5–15 m, and midland areas with elevation ranging from 15–200 m, were ranked the second and third with 26% and 11% weights, respectively. These areas were located to the west of the Delta, including the capital city of Hanoi. Land used for rice production also had the highest contribution to disease occurrence with a 46% weight and followed by land used for built-up purposes at 35%.

**Table 3.** Attribute weight of each group within a factor.

| Factors | Attribute Values of Factors | Attribute Weight (%) |
|---|---|---|
| Water bird density (heads/km$^2$) | 0–892 | 11 |
| | 893–2097 | 76 |
| | 2098–4299 | 7 |
| | >4299 | 6 |
| Chicken density (heads/km$^2$) | 0–1738 | 37 |
| | 1739–3992 | 27 |
| | 3993–9472 | 25 |
| | >9472 | 11 |
| Elevation (m) | ≤5 | 73 |
| | >5–15 | 16 |
| | >15–200 | 11 |
| | >200 | 0 |
| Land use/land cover (%) | Agriculture | 46 |
| | Built-up | 35 |
| | Aquaculture | 17 |
| | Forest/perennial trees | 2 |

The BRT estimation results provided essential information for weighted overlay analysis. The weight of each attribute inTable 3 was assigned to corresponding groups in each raster layer through raster reclassification processes. For instance, in the land-use raster, agriculture, built-up, water, and forest/perennial trees groups were, respectively, assigned their corresponding weight values of 46, 35, 17, and 2 (Table 3). The same procedure was applied to other raster layers detailing water bird density, chicken density, and elevation. Each of the input rasters was then weighed using the weighted values from Table 2. In this weighted overlay analysis, water bird density had a 19% weight, land used for rice paddy field an 18% weight, elevation an 18% weight, land used for aquaculture a 17% weight, land used for built-up purposes a 14% weight, chicken density a 12% weight, and land used for forest/perennial trees a 2% weight. The output suitability map is shown in Figure 3.

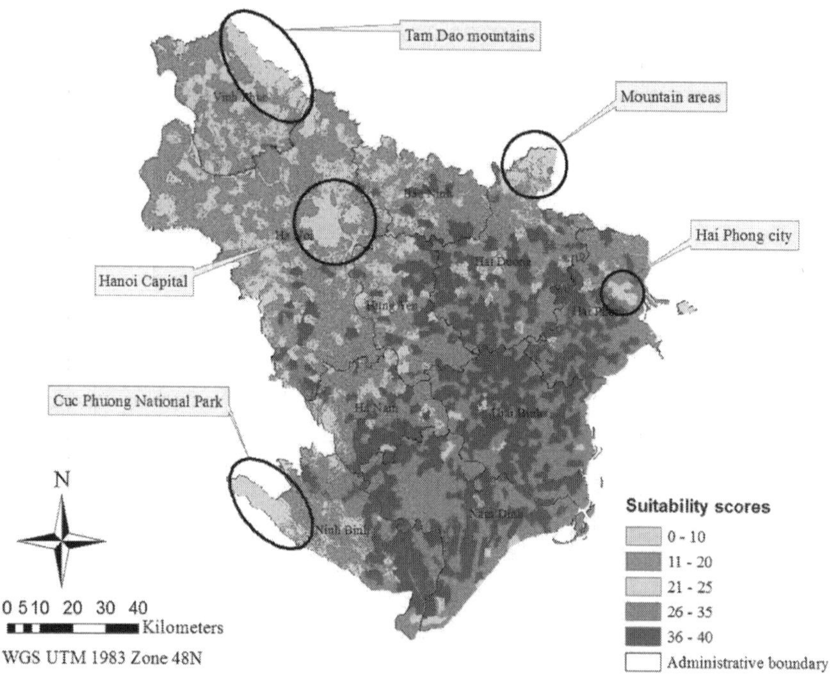


**Figure 3.** Weighted overlay analysis results for the HPAI H5N1 occurrence in the Red River Delta.

The highest suitability score areas were shown in red, followed by orange. Yellow, green, and blue areas have lower suitability scores. These areas (suitability scores ranging from 0–10, 11–20, and 21–25, respectively) were mostly located to the west and northwest of the Red River Delta. The areas with high suitability scores (ranging from 26–40), as shown in red and orange, were chosen as the focus areas for the alternative vaccination program against HPAI H5N1.

### 3.1.2. Cost-Effectiveness Analysis of Vaccination Programs and Policy Implications

The focus areas for the alternative vaccination program were extracted and overlaid with the spatial distribution of the fifth HPAI H5N1 (Figure 4a) for the cost-effectiveness analysis. The fifth wave of outbreak, with 74 affected communes, was the result of the national poultry vaccination program. Based on the Vietnam follow-up disease report No. 45 released by OIE (The World Organisation for Animal Health) for the period from December 2006 to December 2010 [43], we found that the disease was mostly reported in unvaccinated poultry. Vaccines mainly used in Vietnam contain a killed antigen combined with an oil-based adjuvant: (1) A/TK/England/N-28/73, subtype H5N2 (referred to as N28); and (2) a genetically modified reassortant H5N1 low-pathogenic virus, A/Harbin/Re-1/2003 (referred to as Re-1) [44]. Vaccines were found to provide protection against the disease and reduction of viral shedding on both chickens and ducks [44]. The vaccination program proved to be a viable means of protection against the HPAI H5N1 virus [9]. Given this information, this study assumed 100% vaccine efficacy in order to evaluate the tradeoff in cost-effectiveness between the two vaccination programs. As a result, it is expected that the focus areas are protected from the HPAI H5N1 disease as the result of frequent vaccinations under the alternative vaccination program. Figure 4b showed that a total of 61 out of 74 infected communes in the fifth epidemic was correctly predicted in the focus areas for the alternative vaccination program. As a result, the alternative vaccination program would protect these 61 communes from the disease. However, the other 13 communes which were not covered by this program could be affected by the disease.

The cost-effectiveness analysis of the vaccination programs was conducted by using Equations (1) and (2) to investigate which program would be more successful in preventing the disease occurrence. The results of the analysis are shown in Table 4.

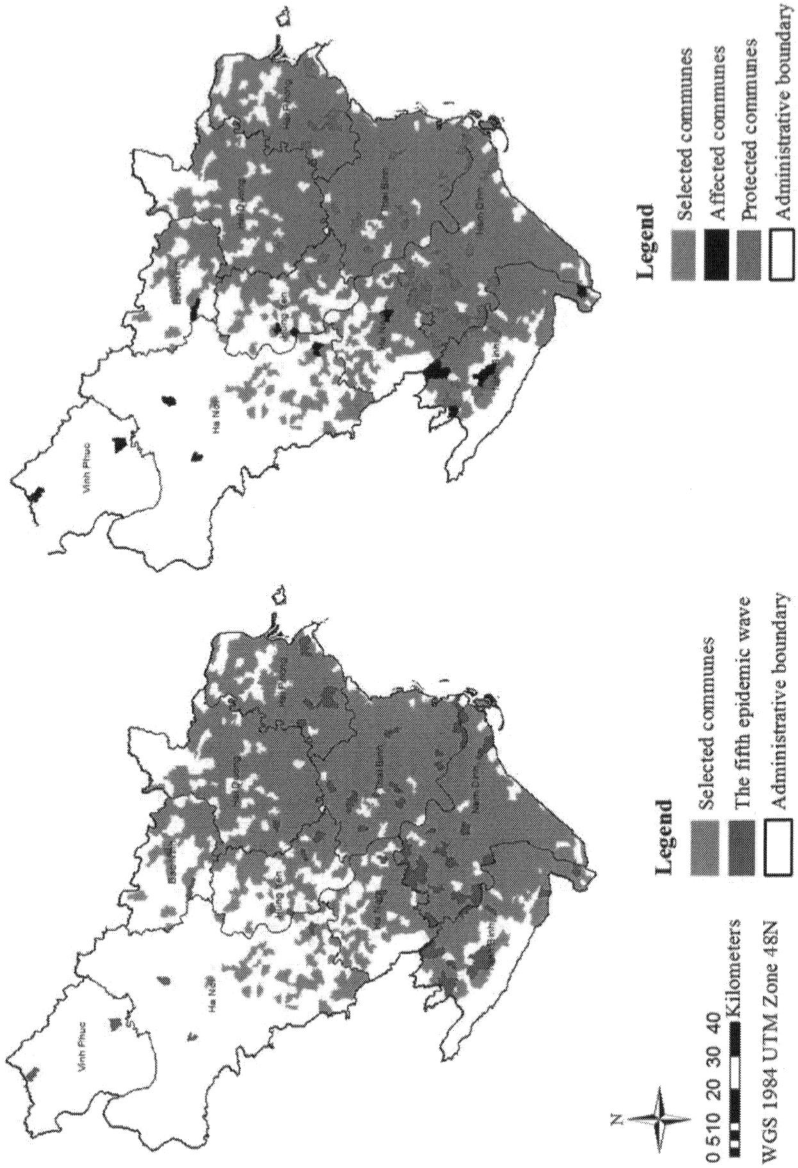


**Figure 4.** (**a**) Spatial distribution of the fifth epidemic wave in the Red River Delta and the focus areas for the alternative vaccination program; and (**b**) spatial distribution of the estimated affected communes and protected communes under the alternative vaccination program.

Of the total number of 2248 communes in the River Delta, there were 193 communes affected by the disease before the implementation of the national poultry vaccination program. This resulted in the infection rate ARV = 8.59. The national poultry vaccination program contributed to the reduction of the affected communes to 74 as reported in the fifth epidemic waves, resulting in the infection rate ARU = 3.29. The alternative vaccination program was expected to further reduce the number of communes affected to 13 communes which yields the infection rate ARU = 0.58. The effectiveness results suggested that the alternative vaccination program would be more successfully in reducing the rate of disease occurrence measured at 93.26%, compared to the national poultry vaccination program at 61.66%.

It was previously estimated that the optimal length of both chicken and duck production cycles were 10 weeks, including a two-week cleaning period [45,46]. Assuming that producers are continuously engaged in production, there would be five duck production cycles per year. Therefore, the alternative vaccination program would involve five vaccination campaigns throughout the year. It was noted that the national poultry vaccination program covered the entire poultry population in the Red River Delta and was conducted twice a year. The cost of a HPAI H5N1 vaccination in Viet Nam was estimated to be U.S. $0.038/head, including the vaccine cost of U.S. $0.016 per dose, labor cost of U.S. $0.013, and other costs associated with vaccination of U.S. $0.009 [47]. As a result, the cost of the national poultry vaccination program was estimated at U.S. $4.50 million per year. The alternative vaccination program consists of about half of the total communes in the Delta, covering 1137 communes with poultry population of 31,171 thousand birds. Having vaccinated poultry five times per year would cost U.S. $5.92 million per year. By examining the costs of vaccination only, the cost of the alternative vaccination program appears to be higher than the national poultry vaccination program (U.S. $5.92 million *vs*. U.S. $4.50 million).

**Table 4.** Cost-effectiveness analysis of the two programs.

| Content | Before Vaccination Program | National Poultry Vaccination Program | Alternative Vaccination Program |
|---|---|---|---|
| Total number of communes involved in vaccination program | 0 | 2248 | 1137 |
| Number of rounds of vaccination per year | 0 | 2 | 5 |
| Number of birds vaccinated per round (Thousand heads) | 0 | 59,241 | 31,171 |
| Number of affected communes | 193 | 74 | 13 |
| Infection rate (%) | 8.59 | 3.29 | 0.58 |
| Effectiveness (%) | | 61.66 | 93.26 |
| Vaccination cost (Million U.S. $) | 0 | 4.50 | 5.92 |
| Number of birds culled (Thousand heads) | 6375 | 2165 | 395 |
| Cost of culling birds (Million U.S. $) | 1.59 | 0.54 | 0.10 |
| Government compensation (Million U.S. $) | 7.95 | 2.68 | 0.49 |
| Number of communes included in emergency vaccination | 543 | 304 | 92 |
| Number of birds vaccinated in emergency vaccination (Thousand heads) | 0 | 8625 | 2898 |
| Cost of emergency vaccination (Million U.S. $) | 0 | 0.33 | 0.11 |
| Farmers loss (Million U.S. $) | 4.95 | 1.65 | 0.30 |
| Total loss (Million U.S. $) | 14.49 | 9.70 | 6.92 |

When the disease occurs, all birds in affected communes are culled due to the stamping out program and all birds in surrounding communes are vaccinated. Before the official vaccination campaign was launched at the end of 2005, the only emergency response to the disease occurrence was the stamping out program. This resulted in 6.375 million. birds in 193 affected communes culled in the second and third epidemic waves. After the implementation of the national poultry vaccination program, a total of 2.165 million birds were affected and culled, and 8.625 million birds were vaccinated, as the result of the emergency response to the disease occurrence. Under the alternative vaccination program, it was estimated that 395,000 birds in 13 communes were affected and culled by the disease. Another 2.898 million birds in 92 surrounding communes were vaccinated due to the emergency vaccination.

The government incurred more losses from the stamping out and emergency vaccination. The cost of culling birds in the stamping out process was estimated at U.S. $0.25/head [47]. Total culling cost was estimated at U.S. $1.59 million before the implementation of the national poultry vaccination program. This cost was much smaller as a result of either the national poultry vaccination program (U.S. $0.54

million) or the proposed alternative vaccination program (U.S. $0.10 million). The average amount of compensation per bird culled due to the disease occurrence was regulated at U.S. $1.24/head (23,000 VND/head) (exchange rate at 1 USD = 18,500 VND), in Decision No 719/QD-TTg, dated 5 June 2008. In addition, the average market value of a bird was estimated at U.S. $2 [12]. The farmers also suffered losses of U.S. $0.76/head from production because of the value difference between the market price and the government compensation. This resulted in the Government's additional losses from compensation and farmers' losses of U.S. $7.95 million and U.S. $4.95 million, respectively, in the second and third epidemic waves. Under the national poultry vaccination program, emergency vaccination was implemented which caused the government an estimated U.S. $0.33 million in losses in addition to the government compensation and farmers losses measured at, respectively, U.S. $2.68 million and U.S. $1.65 million. These losses were also estimated at U.S. $0.49 million, U.S. $0.11 million, and U.S. $0.30 million for the government and farmers under the alternative vaccination program. Except for the costs of vaccination (U.S. $4.50 million *vs.* U.S. $5.92 million), and other costs, the alternative vaccination program costs were lower than the national poultry vaccination program.

Total losses imposed on the government and farmers were highest without vaccination and estimated at U.S. $14.49 million, as compared to U.S. $9.70 million for the national poultry vaccination program, and U.S. $6.92 million for the alternative vaccination program.

## 3.2. Discussion

A vaccination program was identified as a key strategic plan in preventing the HPAI H5N1 disease in Vietnam in the period 2006–2010 [8]. Although the implementation of the national poultry vaccination program resulted in the reduction in disease outbreaks and number of infected birds [44], the disease was repeatedly reported during this period. The third epidemic wave of outbreaks occurred during the first vaccination campaign (September–December 2005). The fourth and fifth waves occurred in between two rounds of vaccination which were from December 2006–March 2007 and May–September 2007, respectively. Infection was reported in unvaccinated ducks based on routine surveillance [10,44]. During the time between the two rounds of vaccinations, the unvaccinated poultry are susceptible to the disease. Additionally, the best vaccination immunity occurs when poultry is vaccinated at the recommended ages suggested by vaccine manufacturers [12]. The national poultry vaccination program scheduled vaccinations at a certain time in each province. It is highly unlikely that the vaccination program exactly matches the optimal timing for vaccination recommended for poultry at various ages.

The alternative vaccination program seeks to fill this gap between the two vaccination rounds and focus more on high risk areas of disease occurrences. The alternative vaccination program was proposed to vaccinate poultry for every production cycle at the recommended poultry ages in high risk areas. Vaccinating poultry at early age provides high level of immunity [44]. A two-dose vaccine given to ducks at one day of age and a booster at four weeks of age produced effective protection [44,48]. It is critical to understand what factors potentially affect the disease occurrence in order to locate susceptible areas where the disease is likely to occur. These factors have been analyzed in previous studies. This study incorporated these factors to spatially identify the focus areas for the proposed alternative vaccination program.

Traditional production methods with free range water bird farming and backyard chicken farming have been considered to be typical Asian production methods which had the potential of contracting and spreading the HPAI H5N1 virus to other neighboring farms [49,50]. These methods were found popular in the poultry sectors 3 and 4 (as classified by [51]) which have small scale production with less than 2000 birds [4]. Poultry sectors 1 and 2, on the other end, are characterized by industrial and commercial poultry production which operate with standard procedures and keep poultry indoors continuously during production to maintain high biosecurity standards against diseasesh, including HPAI H5N1 [51,52]. The BRT results showed the same trend. Water bird density ranging from 893–2097 heads/km$^2$ and chicken density ranging from 0–1738 heads/km$^2$ were found to have higher impacts on the disease occurrence as their weights were estimated at 76% for water bird density factor and 37% in water bird and chicken density factors, respectively. Therefore, it was expected that poultry sectors 3 and 4 would fall more in the lower density group. This result also agreed with the study by [9], which suggested communes with medium water bird density to have an increased risk of contracting the disease. Communes with higher water bird and chicken density would be expected to fall in poultry sectors 1 and 2, which have more secure closed farming methods against the disease than the free range farming.

Lower elevation was previously identified to be correlated with the HPAI H5N1 disease in Vietnam, in general, and in the Red River Delta, Vietnam, in particular [5,17]. This finding was further confirmed by BRT estimation. It was suggested that topographic elevation features noticeably contributed differently to the disease occurrence. Coastal areas with elevation less than 5 m contributed to the likelihood of disease occurrence. In contrast to lower elevation areas, upland areas with elevation greater than 200 m were found not likely to affect the disease. Evergreen forests or forestry production dominates these areas [53]. This was also consistent with BRT estimation for land use/land cover which showed that the weight of land used for forest/perennial trees was small and measured at 2% (see Table 2).

The BRT results for land use/land cover were in agreement with studies by [14] and [17] which suggested the link between HPAI H5N1 occurrence and the higher proportion of land use for rice paddy fields and closer distance to higher-density human population areas. Water bird movement through rice paddy fields has been defined as a potential source for spreading the HPAI H5N1 virus [10,14,16,17,54]. The built-up areas characterizes urban and peri-urban areas. It was found that peri-urban areas were the hotspot for the occurrence of the disease. These are the places where land-use changes, interaction, and contact between human and poultry become more frequent and, therefore, they were found to be at significantly higher risk for HPAI H5N1 occurrence [19].

The suitable areas for the vaccination campaign are shown in Figure 3. These suggested areas are characterized by low and high suitability scores. The lowest suitability score areas were either in urban cores or mountainous areas. They included urban core areas of Hanoi, Hai Phong, Hai Duong, Bac Ninh, Hung Yen, Nam Dinh, and Thai Binh provinces and mountain areas of Ba Vi of Hanoi, Tam Dao of Vinh Phuc, Cuc Phuong national park of Ninh Binh, and mountain areas located to the north of the Chi Linh district of Hai Duong province. The highest suitability score areas were mostly located in the coastal areas to the east and south of Hanoi. These areas were chosen to have frequent vaccinations against the HPAI H5N1 disease in the Red River Delta. A total of 1137 communes, corresponding to 50.6% of total communes in the Delta, were selected for the alternative vaccination program, including provinces near the Gulf of Tonkin—Hai Phong, Thai Binh, Nam Dinh, Hai Duong, and the eastern parts of Hung Yen and Ha Nam provinces. These areas were also previously identified to have the highest probability of disease occurrence in the Delta [5]. Almost the entire areas of Hai Duong, Hai Phong, Thai Binh, and Nam Dinh provinces were identified as the focus areas for the alternative vaccination program except urban cores and mountains in the north of Hai Duong. Water bird production with free range farming was found to be the most intensive in the Red River Delta together with the Mekong Delta [55]. Thanh Oai, Thuong Tin, Ung Hoa, and Phu Xuyen districts of Hanoi were also identified as focus areas. These areas are famous for high-quality free-range duck meat provided to consumers in Hanoi.

The cost-effectiveness analysis indicated that the alternative vaccination program would better prevent the occurrence of the disease at lower costs associated with disease prevention than the national poultry vaccination program. Although the cost of the alternative vaccination program was estimated to be higher than that of the national poultry vaccination program (U.S. $5.92 million *vs*. U.S. $4.50 million), it resulted in reducing the rate of disease occurrence by 93.26% as compared to 61.66% of the national poultry vaccination program. As a result, this reduced the cost of immediate responses to the disease occurrence, including stamping out and emergency vaccination, government compensation, and farmers' losses. This contributed to lower total costs associated with the alternative vaccination program as compared to the national poultry vaccination program (U.S. $6.92 million *vs.* U.S. $9.70). The results of the analysis suggested that Vietnam may face lower costs with the alternative vaccination program.

Poultry production is much more complicated in reality. Producers may have different production cycles. Therefore, for the implementation of the alternative vaccination program, it is recommended that vaccination occurs at the commune level where the commune veterinary officers are required to monitor and vaccinate all poultry at recommended ages for every production cycle.

## 4. CONCLUSIONS

The national poultry vaccination program in the period 2006–2010, which implemented an annual two round vaccination plan for the entire geographical area of the Red River Delta, did not successfully control the disease. This study explored implications of an alternative vaccination program. This alternative vaccination program involves vaccination for every production cycle at the recommended poultry age in high risk areas. This alternative plan would have to be enacted at the local level for all production cycles.

This study identified the focus areas for the alternative vaccination program which were located mostly in the coastal areas to the east and south of Hanoi. A total of 1137 communes, corresponding to 50.6% of total communes in the Delta, were selected for the alternative vaccination program. The alternative vaccination program would have been less costly as compared to the national poultry vaccination program. Effectiveness analysis found that the alternative vaccination program would have been more successful in reducing the rate of disease occurrence from 61.66% rate of reduction (the national poultry vaccination program) to 93.26% rate of reduction (the alternative vaccination program). The cost analysis indicated that the alternative vaccination program would have saved the government and farmer resources because of lower total costs associated with prevention. Total losses imposed on both the government and farmers were higher for the national poultry vaccination program (U.S. $9.70 million) than for the alternative vaccination program (U.S. $6.92 million).

## ACKNOWLEDGMENTS

This study is a contribution to the project "Coupled Natural-Human Systems and Emerging Infectious Diseases: Anthropogenic environmental change and avian influenza in Vietnam", funded by the U.S. National Science Foundation Grant # DEB-0909410. We would like to thank the East-West Center for providing computer laboratory facilities and support. Thanks to the Vietnam Department of Animal Health and the General Statistic Office of Vietnam for providing data for the study.

## AUTHOR CONTRIBUTIONS

All the authors contributed to this study are as follows: Chinh C. Tran designed, carried out the analysis, interpreted the results and wrote this manuscript. John F. Yanagida supervised the methods of statistical and geostatistical analysis and contributed to the design and revision of the article. Sumeet Saksena provided advice for the Boosted Regression Trees model and reviewed the manuscript for intellectual content. Jefferson Fox reviewed the manuscript for intellectual content and contributed to the revision of the article. All the authors read and approved the final manuscript.

## CONFLICTS OF INTEREST

The authors declare no conflict of interest.

# REFERENCES


1 OIE. Follow-Up Report No. 8. Available online: http://www.oie.int/wahis_2/temp/reports/en_fup_0000014834_20140225_152400.pdf (accessed on 5 September 2014).

2 Phan Dang, T.; Duquesne, B.; Lebailly, P.; Vu Dinh, T. Diversification and epidemic risks of poultry production systems in Hanoi suburban. *J. Sci. Dev.* **2010**, *8*, 203–215.

3 Peyre, M.; Desvaux, S.; Phan Dang, T.; Rossi, V.; Renard, J.F.; Vu Dinh, T.; Roger, F. Financial evaluation of vaccination strategies against HPAI. A modeling approach. In Proceedings of AI Research to Policy International Workshop, FAO, Hanoi, Vietnam, 16–18 June 2008.

4 Desvaux, S.; Ton, V.D.; Phan Dang, T.; Hoa, P.T.T. *A General Review and Description of the Poultry Production in Vietnam*; Agricultural Publishing House: Hanoi, Vietnam, 2008.

5 Tran, C.C.; Yost, R.S.; Yanagida, J.F.; Saksena, S.; Fox, J.; Sultana, N. Spatio-temporal occurrence modeling of highly Pathogenic Avian Influenza subtype H5N1: A case study in the Red River Delta, Vietnam. *ISPRS Int. J. Geo-Inf.* **2013**,*2*, 1106–1121.

6 World Bank. *The Avian Influenza Emergency Recovery Project: Implementation completion and Results Report*; World Bank: Washington, DC, USA, 2007.

7 CAP. *Policy Analysis of HPAI Strategy Including Analysis of Collaboration and Partnership between Public and Private Sectors: Gathering Evidence for A Transitional Strategy (GETS) for HPAI H5N1 Vaccination in Vietnam*; Center for Agricultural Policy: Hanoi, Vietnam, 2011.

8 MARD & MOH. *Vietnam: Integrated National Operational Program for Avian and Human Influenza (OPI) 2006–2010*; MARD & MOH: Hanoi, Vietnam, 2006.

9 Henning, K.A.; Henning, J.; Morton, J.; Long, N.T.; Ha, N.T.; Meers, J. Farm-and flock-level risk factors associated with Highly Pathogenic Avian Influenza outbreaks on small holder duck and chicken farms in the Mekong Delta of Viet Nam. *Prev. Vet. Med.* **2009**, *91*, 179–188.

10 Minh, P.Q.; Morris, R.S.; Schauer, B.; Stevenson, M.; Benschop, J.; Nam, H.V.; Jackson, R. Spatio-temporal epidemiology of highly Pathogenic Avian Influenza outbreaks in the two deltas of Vietnam during 2003–2007. *Prev. Vet. Med.* **2009**, *89*, 16–24.

11 Taylor, N.; Dung, D.H. *An Analysis of Data Generated by Post-Vaccination Sero-Monitoring and Surveillance Activities, Following HPAI Vaccination in Viet Nam (2005–2006)*; FAO: Hanoi, Vietnam, 2007; p. 46.

12 Sims, L.; Do, H.D. *Vaccination of Poultry in Vietnam against H5N1 Highly Pathogenic Avian Influenza*; Australian Government Department of Agriculture, Fisheries and Forestry: Canberra, Australia, 2009.

13 MARD & MOH. Avian and pandemic influenza Vietnam's experience. In Proceedings of International Ministerial Conference on Animal and Pandemic Influenza, Hanoi, Vietnam, 19–21 April 2010.

14 Pfeiffer, D.U.; Minh, P.Q.; Martin, V.; Epprecht, M.; Otte, M.J. An analysis of the spatial and temporal patterns of highly Pathogenic Avian Influenza occurrence in Vietnam using national surveillance data. *Vet. J.* **2007**, *174*, 302–309.

15 Desvaux, S.; Grosbois, V.; Pham, T.T.H.; Fenwick, S.; Tollis, S.; Pham, N.H.; Tran, A.; Roger, F. Risk factors of highly Pathogenic Avian Influenza H5N1 occurrence at the village and farm levels in the Red River Delta Region in Vietnam.*Transbound. Emerg. Dis.* **2011**, *58*, 492–502.

16 Gilbert, M.; Chaitaweesup, P.; Parakamawongsa, T.; Premashthira, S.; Tiensin, T.; Kalpravidh, W.; Wagner, H.; Slingenbergh, J. Free-grazing ducks and highly Pathogenic Avian Influenza, Thailand. *Emerg. Infect. Dis.* **2006**, *12*, 227–234.

17 Gilbert, M.; Xiao, X.; Pfeiffer, D.U.; Epprecht, M.; Boles, S.; Czarnecki, C.; Chaitaweesub, P.; Kalpravidh, W.; Minh, P.Q.; Otte, M.J.; *et al.* Mapping H5N1 highly Pathogenic Avian Influenza risk in southeast Asia. *Proc. Natl. Acad. Sci. USA* **2008**, *105*, 4769–4774.

18 FAO Summary of Highly Pathogenic Avian Influenza (HPAI) Situation in Viet Nam 31 December 2008. Available online: http://www.un.org.vn/images/stories/press_centre/2008_hpai_update_31dec08.pdf (accessed on 21 October 2010).

19 Saksena, S.; Fox, J.; Epprecht, M.; Tran, C.; Nong, D.; Spencer, J.; Nguyen, L.; Finucane, M.; Tran, V.; Wilcox, B. Evidence for the convergence model: The emergence of highly Pathogenic Avian Influenza (H5N1) in Viet Nam. *PLoS ONE* **2015**, *10*, e0138138.

20 CGIAR-CSI. Shuttle Radar Topography Mission (SRTM) 90-m Resolution Digital Elevation Model (DEM). Available online: http://srtm.csi.cgiar.org/ (accessed on 2 March 2015).

21 USGS. USGS EROS Data Center. Available online: http://glovis.usgs.gov/ (accessed on 28 October 2013).

22 ESRI. How Weighted Overlay Works. Available online: http://resources.arcgis.com/en/help/main/10.1/index.html#//009z000000s1000000 (accessed on 10 September 2014).

23 Gilbert, M.; Xiao, X.; Domenech, J.; Lubroth, J.; Martin, V.; Slingenbergh, J. Anatidae migration in the western Palearctic and spread of highly Pathogenic Avian Influenza H5N1 virus. *Emerg. Infect. Dis.* **2006**, *12*, 1650–1656.

24 Shahid, S. Impact of climate change on irrigation water demand of dry season Boro rice in northwest Bangladesh. *Clim. Change* **2011**, *105*, 433–453.

25 Jayakumar, S.; Arockiasamy, D.I.; Britto, S.J. Conserving forests in the eastern Ghats through remote sensing and GIS-A case study in Kolli hills. *Curr. Sci.* **2002**, *82*, 1259–1266.

26 Münch, Z.; Conrad, J. Remote sensing and GIS based determination of groundwater dependent ecosystems in the Western Cape, South Africa. *Hydrogeol. J.* **2007**, *15*, 19–28.

27 Diamond, J.T.; Wright, J.R. Design of an integrated spatial information system for multiobjective land-use planning.*Environ. Plan. B-Plan. Design* **1988**, *15*, 205–214.

28 Elith, J.; Leathwick, J.R.; Hastie, T. A working guide to boosted regression trees. *J. Anim. Ecol.* **2008**, *77*, 802–813.

29 Elith, J.; Catherine, G.; Anderson, P.R.; Dudík, M.; Ferrier, S.; Guisan, A.; Hijmans, J.R.; Huettmann, F.; Leathwick, R.J.; Lehmann, A.; *et al*. Novel methods improve prediction of species' distributions from occurrence data. *Ecography***2006**, *29*, 129–151.

30 Martin, V.; Pfeiffer, D.U.; Zhou, X.; Xiao, X.; Prosser, D.J.; Guo, F.; Gilbert, M. Spatial distribution and risk factors of highly Pathogenic Avian Influenza (HPAI) H5N1 in China. *PLoS Pathog.* **2011**, *7*, e1001308.

31 Gilbert, M.; Newman, S.H.; Takekawa, J.Y.; Loth, L.; Biradar, C.; Prosser, D.J.; Balachandran, S.; Rao, M.V.S.; Mundkur, T.; Yan, B.; *et al*. Flying over an infected landscape: Distribution of highly Pathogenic Avian Influenza H5N1 risk in south Asia and satellite tracking of wild waterfowl. *Ecohealth* **2010**, *7*, 448–458.

32 Stevens, K.B.; Pfeiffer, D.U. Spatial modelling of disease using data-and knowledge-driven approaches. *Spat. Spatio-Temporal Epidemiol.* **2011**, *2*, 125–133

33 Van Boeckel, T.P.; Thanapongtharm, W.; Robinson, T.; Biradar, C.M.; Xiao, X.; Gilbert, M. Improving risk models for avian influenza: The role of intensive poultry farming and flooded land during the 2004 Thailand epidemic. *PLoS ONE* **2012**, *7*, e49528.

34 Jenks, G.F. The data model concept in statistical mapping. *Int. Yearb. Cartogr.* **1967**, *7*, 186–190.

35 Castrence, M.; Nong, D.H.; Tran, C.C.; Young, L.; Fox, J. Mapping urban transitions using multi-temporal Landsat and

DMSP-OLS night-time lights imagery of the Red River Delta in Vietnam. *Land* **2014**, *3*, 148–166.

36 Schneider, A. Monitoring land cover change in urban and peri-urban areas using dense time stacks of Landsat satellite data and a data mining approach. *Remote Sens. Environ.* **2012**, *124*, 689–704.

37 Mountrakis, G.; Im, J.; Ogole, C. Support vector machines in remote sensing: A review. *ISPRS J. Photogramm. Remote Sens.* **2011**, *66*, 247–259.

38 Weinberg, G.A.; Szilagyi, P.G. Vaccine epidemiology: Efficacy, effectiveness, and the translational research roadmap.*J. Infect. Dis.* **2010**, *201*, 1607–1610.

39 Pfeiffer, D.U. Assessment of H5N1 HPAI risk and the importance of wild birds. *J. Wildl. Dis.* **2007**, *43*, S47–S50.

40 Smith, G.J.D.; Fan, X.H.; Wang, J.; Li, K.S.; Qin, K.; Zhang, J.X.; Vijaykrishna, D.; Cheung, C.L.; Huang, K.; Rayner, J.M.; *et al.* Emergence and predominance of an H5N1 influenza variant in China. *Proc. Natl. Acad. Sci. USA* **2006**, *103*, 16936–16941.

41 Songserm, T.; Jam-on, R.; Sae-Heng, N.; Meemak, N.; Hulse-Post, D.J.; Sturm-Ramirez, K.M.; Webster, R.G. Domestic ducks and H5N1 influenza epidemic, Thailand. *Emerg. Infect. Dis.* **2006**, *12*, 575.

42 Webster, R.G.; Hulse-Post, D.J.; Sturm-Ramirez, K.M.; Guan, Y.; Peiris, M.; Smith, G.; Chen, H. Changing epidemiology and ecology of highly pathogenic avian H5N1 influenza viruses. *Avian Dis.* **2007**, *51*, 269–272.

43 OIE. Follow-Up Report No. 45. Available online: http://web.oie.int/wahis/reports/en_fup_0000010005_20101204_123300.pdf (accessed on 15 August 2014).

44 Pfeiffer, J.; Suarez, D.L.; Sarmento, L.; To, T.L.; Nguyen, T.; Pantin-Jackwood, M.J. Efficacy of commercial vaccines in protecting chickens and ducks against H5N1 highly Pathogenic Avian Influenza viruses from Vietnam. *Avian Dis.***2010**, *54*, 262–271.

45 Tran, C.C. *Public Policy Instruments for Risk Management of Highly Pathogenic Avian Influenza Subtype H5N1 in Vietnam*; University of Kentucky: Lexington, KY, USA, 2010.

46 Tran, C.C.; Yanagida, J.F. Computational economic analysis of duck production at the farm household level in the context of highly Pathogenic Avian Influenza subtype H5N1 in the Red River Delta, Vietnam. *Asian J. Agric. Ext. Econ. Sociol.* **2015**, *6*, 172–184.

47 Hinrichs, J.; Sims, L.; McLeod, A. Some direct costs of control for avian influenza. In Proceedings of the 11th International Society for Veterinary Epidemiology and Economics, Cairns, Australia, August 2006.

48 Beato, M.S.; Toffan, A.; de Nardi, R.; Cristalli, A.; Terregino, C.; Cattoli, G.; Capua, I. A conventional, inactivated oil emulsion vaccine suppresses shedding and prevents viral meat colonisation in commercial (Pekin) ducks challenged with HPAI H5N1. *Vaccine* **2007**, *25*, 4064–4072.

49 Alhaji, N.B.; Odetokun, I.A. Assessment of biosecurity measures against highly Pathogenic Avian Influenza risks in small-scale commercial farms and free-range poultry flocks in the northcentral Nigeria. *Transbound. Emerg. Dis.* **2011**,*58*, 157–161.

50 FAO. Bio-Security for Highly Pathogenic Avian Influenza: Issues and Options. Available online: ftp://ftp.fao.org/ docrep/fao/011/ i0359e/i0359e00.pdf (accessed on 26 May 2012).

51 FAO. FAO Recommendations on the Prevention, Control and Eradication of Highly Pathogenic Avian Influenza (HPAI) in Asia. Available online: http://web.oie.int/eng/ AVIAN_INFLUENZA/ FAO recommendations on HPAI.pdf(accessed on 26 May 2012).

52 FAO. The Importance of Biosecurity in Reducing HPAI Risk on Farms and in Markets. Available online: http:// www.fao.org/docs/eims/upload//236621/ah691e.pdf (accessed on 26 May 2012).

53 EAP-AP. Land Cover Assessment and Monitoring. Available online: http://www.rrcap.ait.asia/lc/cd/html/vietnam.html (accessed on 8 January 2015).

54 Gilbert, M.; Xiao, X.; Chaitaweesub, P.; Kalpravidh, W.; Premashthira, S.; Boles, S.; Slingenbergh, J. Avian influenza, domestic ducks and rice agriculture in Thailand. *Agric. Ecosyst. Environ.* **2007**, *119*, 409–415.

55 Edan, M. *Review of Free-Range Duck Farming Systems in Northern Vietnam and Assessment of Their Implication in The Spreading of The Highly Pathogenic (H5N1) Strain of Avian Influenza (HPAI)*; Agronomes et Veterinaires sans Frontieres: Lyon, France, 2006; p. 101.

# CHAPTER 6

# Can Probiotics Improve the Environmental Microbiome and Resistome of Commercial Poultry Production?

**Adriana A. Pedroso,[1] Anne L. Hurley-Bacon,[2] Andrea S. Zedek,[3] Tiffany W. Kwan,[1] Andrea P. O. Jordan,[4] Gloria Avellaneda,[5] Charles L. Hofacre,[1] Brian B. Oakley,[6] Stephen R. Collett,[1] John J. Maurer,[1] and Margie D. Lee[1,*]**

1 Poultry Diagnostic and Research Center/Center for Food Safety, The University of Georgia, Athens, GA 30602, USA;

2 Merial, Athens, GA 30601, USA

3 Zoetis Animal Health, Madison, NJ 07932, USA

4 Perelman School of Medicine, University of Pennsylvania, Philadelphia, PA 19104, USA;

5 Loehmann Animal Health International, Waterville, ME 0491, USA;

6 Richard B. Russell Agricultural Research Center, Agricultural Research Service, South Atlantic Area, Athens, GA 30605, USA

## ABSTRACT

Food animal production systems have become more consolidated and integrated, producing large, concentrated animal populations and significant amounts of fecal waste. Increasing use of manure and litter as a more "natural" and affordable source of fertilizer may be contributing to contamination of fruits and vegetables with foodborne pathogens. In addition, human and animal manure have been identified as a significant source of antibiotic resistance genes thereby serving as a disseminator of resistance to soil and waterways. Therefore, identifying methods to remediate human and animal waste is critical in developing strategies to improve food safety and minimize the dissemination of antibiotic resistant bacteria. In this study, we sought to determine whether withdrawing antibiotic growth promoters or using alternatives to antibiotics would reduce the abundance of antibiotic resistance genes or prevalence of pathogens in poultry litter. Terminal restriction fragment length polymorphism (T-RFLP) paired with high throughput sequencing was used to evaluate the bacterial community composition of litter from broiler chickens that were treated with streptogramin growth-promoting antibiotics, probiotics, or prebiotics. The prevalence of resistance genes and pathogens was determined from sequencing results or PCR screens of litter community DNA. Streptogramin antibiotic usage did not elicit statistically significant differences in Shannon diversity indices or correlation coefficients among the flocks. However, T-RFLP revealed that there were inter-farm differences in the litter composition that was independent of antibiotic usage. The litter from all farms, regardless of antibiotic usage, contained streptogramin resistance genes (*vatA*, *vatB*, and *vatE*), macrolide-lincosamide-streptogramin B resistance genes (*ermA* and *ermB*), the tetracycline resistance gene *tetM* and class 1 integrons. There was inter-farm variability in the distribution of *vatA* and *vatE* with no statistically significant differences with regards to usage. Bacterial diversity was higher in litter when probiotics or prebiotics were administered to flocks but as the litter aged, diversity decreased. No statistically signficant differences were detected in the abundance of class 1 integrons where 3%–5% of the community was estimated to harbor a copy. Abundance of pathogenic *Clostridium* species increased in aging litter despite the treatment while the abundance of tetracycline-resistant coliforms was unaffected by treatment. However some treatments decreased the prevalence of *Salmonella*. These findings suggest that withdrawing antibiotics or administering alternatives to antibiotics can change the litter bacterial community and reduce the prevalence of some pathogenic bacteria, but may not immediately impact the prevalence of antibiotic resistance.

antibiotic, growth promoter, microbiota, litter, probiotic, prebiotic, streptogramin, integrin

# 1. INTRODUCTION

Modern American agriculture has become successful in producing food security despite having less than 1% of the US population actually listing farming as their occupation [1]. The US Census of Agriculture estimates that 8.6 billion broiler chickens, 107 million turkeys, 96 million beef cattle, and 68 million pigs were sold for meat in 2007 with an additional 350 million laying hens and 79 million dairy cows producing eggs and milk [1]. Eighteen to 36 kg of manure are produced per day per 454 kg animal unit [2] resulting in new challenges in manure management for farmers. The United States Department of Agriculture (USDA) estimates that food animals produce 360 million tons of dry matter per year [3]. Because of increasing costs of chemical fertilizers, manure and poultry litter have become marketable as organic fertilizers. Currently the USDA allows manures to be used as fertilizer to grow organic fruits, vegetables, and grains meant for human consumption as long as specific guidelines are followed for land application and composting [4]. The US organic market has been rapidly growing, with organic food sales accounting for over 3% of total US food sales in 2008 and now it exceeds $20 billion annually in value [5].

Animal manure has been long known to contain varying levels of zoonotic pathogens such as *Salmonella*,*Clostridium* spp., *E. coli*, and *Campylobacter*, among others. Composting can however reduce the levels significantly [6,7], but composting of animal manures and their dissemination are much less regulated than that of human sewage biosolids [8]. The potential ecological impacts of large scale use of animal manures for land management were recently reviewed, which revealed some new concerns regarding antibiotic resistance [9]. The common perspective is that antimicrobial usage in animal production is a significant contributor to the dissemination of resistant zoonotic pathogens such as *Salmonella*, *Campylobacter* and *Staphylococcus*[10]. A number of studies have also found that manure, poultry litter, and contaminated soils contain a high abundance of antibiotic resistance genes, some of which are encoded by class 1 integrons [11,12,13,14,15,16,17,18,19]. In particular poultry litter has been shown to contain an inexplicably high abundance of resistance genes for antibiotics that are not currently used in the industry suggesting that early usage may have had an amplifying effect that has not been overcome by reducing the selective pressure for resistance [16,20,21].

The gastrointestinal tract is home to a diverse population of microorganisms that can have a great impact on host health. In a bird's natural setting, gastrointestinal organisms are rapidly obtained from the egg shell and by consuming fecal material from the adult hen [22]. In today's commercial poultry industry however, the development of a complex intestinal microbiota is delayed because, in order to reduce disease transmission, commercial hatcheries have severed the connection between the chick and the complex fecal microbiota associated with the hen. Effective application of hygiene at the hatchery allows environmental bacteria to play a larger role in seeding the intestinal tract of the commercial broiler chick [22,23]. This becomes important when you consider that the majority of the United States commercial poultry industry uses a built up litter system, meaning that the bacteria left in the litter from the previous flock(s) could potentially have a major impact, positive or negative, on the gastrointestinal integrity of the young broiler chick. Historically, commercial poultry producers have fed low levels of antibiotics in their poultry ration to negate the negative impact of pathogen growth and enteric disease on feed conversion ratio and weight gain. Virginiamycin (streptogramin A and B) is an antibiotic that has been used in agriculture for almost 20 years. Its main use has been as a growth-promoting agent in animal feed in United States and Europe and to prevent necrotic enteritis caused in chickens by *Clostridium perfringens* [24]. This antibiotic and five others were banned in the European Union in 1999 amid concern at the cross-resistances to the streptogramins used for human therapy, dalfo-/quinuprinstin [24,25]. In the US, many poultry companies are establishing antibiotic-free programs for their production system [26,27] utilizing alternatives for disease control.

Probiotics, competitive exclusion products and prebiotics are all antibiotic alternatives that have been shown, in some studies, to prevent the establishment of pathogens in the intestinal tract of chickens, thus increasing weight gain, feed conversion ratio, and livability [27,28,29,30,31]. Probiotics have been defined as live microbial feed supplements designed to benefit the host by improving the intestinal microbial ecology [32]. Prebiotics are defined as a non-digestible food ingredient that beneficially affects the host by selectively stimulating the growth and/or activity of one or a limited number of bacteria in the intestine [33]. Prebiotics utilize complex carbohydrates that serve as nutrients for beneficial bacteria, competitive binding sites to mediate pathogen passage through the intestine, or immune modulators that reduce inflammation or stimulate the mucosal turnover rate [33,34].

While there have been many studies focusing on the effects of antibiotic alternatives on the microbial ecology of the poultry intestine [26,35,36,37,38,39,40,41,42,43,44,45], few studies have characterized the poultry litter microbiota by molecular ecology techniques [19,46]. The objectives of this study were to determine the microbial composition of the litter over multiple production cycles and the impact of alternatives to antibiotics on the ecology and the antibiotic resistome of broiler litter.

# 2. MATERIAL AND METHODS

## 2.1. Housing, Litter and Treatments for Birds Raised on Commercial Poultry Farms

A commercial poultry company in North Georgia (USA), that uses the phosphoglycolipid antibiotic, flavomycin, as the antibiotic growth promoter (AGP) on its contract broiler chicken farms agreed to replace flavomycin with virginiamycin (streptogramin antibiotic) in the finisher feed for three successive flocks. Four houses on each of three poultry farms were enrolled in the study. The birds in two houses were administered virginiamycin for three flocks, and two houses, designated as controls, did not receive any antibiotics during the experimental period (Table 1). For houses receiving AGP, flavomycin (2.2 g/T of feed) was used in the starter, and grower feed for all flocks but only in the finisher feed for the first and fifth flocks. Virginiamycin (22 g/T of feed) was used in the finisher feed, which is provided for the last two weeks of the flocks' growth period, for flocks 2–4. Five successive flocks on the three farms were sampled over a 12 month period. Litter samples (n = 3) from different locations in each house were collected near the end of the grow-out when the broiler chickens were seven weeks of age.

**Table 1** Experimental design and treatments for broiler chickens raised on 3 commercial poultry farms. Two houses on each farm received no antibiotics; two houses received flavomycin (2.2 g/T) in the starter, and grower feed and either flavomycin (2.2 g/T) or ...

| House | Flock 1 | Flock 2 | Flock 3 | Flock 4 | Flock 5 |
|---|---|---|---|---|---|
| A and B | No AGP | No AGP | No AGP | No AGP | No AGP |
| C and D | | | | | |
| Starter- | Flavomycin | Flavomycin | Flavomycin | Flavomycin | Flavomycin |
| Grower- | Flavomycin | Flavomycin | Flavomycin | Flavomycin | Flavomycin |
| Finisher- | Flavomycin | Virginiamycin | Virginiamycin | Virginiamycin | Flavomycin |

## 2.2. Housing, Litter and Treatments for Birds Raised in Research Facilities

Five 9.29 m$^2$ colony houses equipped with fresh pine shaving litter were top-dressed with 75.71 L/house of built-up litter obtained from a Northeast Georgia commercial broiler house. Five hundred day-of-hatch broiler chicks were obtained from a commercial hatchery, and 100 chicks were randomly allocated to each of the colony houses at a stocking density of 930 cm$^2$/bird. Birds were grown to approximately 4–5 weeks of age and received All-Lac, the combination of All-Lac and Biomos, Aviguard or Primalac; the control group did not receive any treatments. Product description and dosaging are described in Table 2. At the end of the growth period, birds were removed and the litter was left in the house for 2–3 weeks before a new flock was placed on the used litter. Each colony house was repopulated with day-old chicks from the same commercial hatchery, the designated probiotic product was applied for each treatment group again and this cycle was repeated for a total of four production cycles. Weekly, five random litter samples from each colony house were obtained using a number/grid system and pooled. A 5 g sub-sample was then taken from the pool and mixed with 20 mL of 0.9% NaCl, homogenized for two minutes and stored at −80 °C for further analysis.

**Table 2** Products and treatment protocols for broiler chickens raised in research facilities.

| Product (Manufacturer) | Composition | Dosage and administration |
|---|---|---|
| All-Lac (Alltech Inc., Lexington, KY, USA) | *Lactobacillus*, *Enterococcus*, and *Pediococcus* | 5 g in 600 mL water for 2,000 birds |
| All-Lac + BioMos (Alltech Inc., Lexington, KY, USA) | All-Lac plus a mannan oligosaccharide derived from the cell wall of *Saccharomyces cerevisiae* | All-Lac: 5 g in 600 mL water for 2,000 birds, BioMos: 2 kg/T of starter feed for 10 days, 1 kg/T of grower feed to 21 days, 0.5 kg/T of finisher feed to 35 days |
| Aviguard (Microbial Developments Ltd, Malvern, UK) | Undefined bacteria collected and cultured from chicken cecum contents | 1 pack in 1,000 mL water for 2,000 birds |
| Primalac (Star-Labs, Clarksdale, MO, USA) | *Lactobacillus* species, *Enterococcus faecium*, and *Bifidobacterium thermophilum* | 1 kg/T of starter to 10 days, 1 kg /T of grower to 21 days, 0.5 kg/T of finisher to 35 days |

## 2.3. Quantifying Tetracycline-Resistant Coliforms in Litter and Fecal Samples

Each week, a sample was taken from the pooled sample described above and cultured to detect coliforms. In brief, 5 g was mixed with 20 mL of sterile saline solution. Samples were then serially diluted in saline and plated in triplicate onto a MacConkey agar plate and MacConkey containing 10 μg/mL tetracycline. The plates were incubated overnight at 37 °C and enumerated after 18 h by counting colony forming units (CFU).

At the end of each cycle, 30 fecal samples were obtained from each flock. For 10 fecal samples, dilutions were made and plated as described above in order to quantify the *E. coli*. In addition, one hundred mg of each of 20 fecal samples was added to 0.9 mL of EC broth containing 10 μg/mL tetracycline in order to detect low levels of resistant *E. coli*. These were incubated at room temperature for 18 h then 10 μL plated on

MacConkey agar plate containing tetracycline in order to detect low levels of resistant coliforms.

## 2.4. Collection of the Bacterial Pellet

The protocol used for isolating bacteria from litter samples was modified from a previously published protocol [47]. Fifteen grams of each sample was suspended in 50 mL of sodium phosphate buffer (pH 8), containing 0.1% Tween 80. The suspensions were mixed vigorously for one minute and the contents poured into a filtration column consisting of prewetted crumpled gauze and cotton in a 60 mL syringe barrel placed within a 50 mL conical tube. A pellet was obtained after centrifugation at 4,500 × g for 20 min at 4 °C. The pellet was resuspended in 1 mL of sterile saline (0.9% NaCl solution in water) and centrifuged at 7,500 × g for 10 min. The resulting bacterial pellet was resuspended in sterile freezer stock medium (1% peptone, 15% glycerol in water) at a 1:10 weight:volume ratio. The samples were stored at −20 °C until processing.

## 2.5. DNA Extraction

Bacterial cells were lysed using beads, solution 1 and IRS of Mo Bio Soil DNA extraction kit (Mo Bio Laboratories Inc., Carlsbad, CA, USA) by vortexing at maximum speed for 40 min [19]. Lysed cells were treated with sodium dodecyl sulfate (final concentration, 0.5%) and proteinase K (final concentration, 0.1 μg/mL) and incubated at 37 °C for 30 min. The sample was extracted twice with an equal volume of phenol-chloroform-isoamyl alcohol (25:24:1) and once with chloroform-isoamyl alcohol (24:1). 20 μL of 10 μg/mL RNAse (DNAse-free) was added to each sample and incubated at 37 °C for 15 min. DNA was concentrated with a 0.6 volume of isopropanol and resuspended in sterile water. The presence of DNA was verified by agarose gel electrophoresis.

## 2.6. PCR Detection of Antibiotic Resistance Genes and Salmonella

DNA extracted from litter was used in PCR to detect the presence of macrolide-lincosamide-streptogramin B (*ermA-C*), streptogramin A (*vatA,B,D,E*, *vgaB*), tetracycline resistance genes (*tetM*, *tetO*) and class 1 integrase (*intI1*) and *Salmonella* as previously described [48,49].

*intI1* abundance was determined using quantitative real-time PCR as described by Nandi *et al.* [16]. In brief, an *E. coli* strain harboring R100-1 (GenBank accession AP000342) was used to normalize the quantitative PCR signal. The *E. coli* R100-1 strain contains one plasmid copy of *intI1* and 7 genomic copies of 16S rRNA. A standard curve relating the cycle threshold (Ct) value to gene copy number was determined for every experiment. Two replicate experiments were performed using DNA extracted from the last litter sample for each flock; replicates were acceptable if the R2 values were 0.98 or greater for both *intI1* and 16S rRNA primer sets. The number of bacteria per PCR reaction was determined by *E. coli* plate counts enabling gene copy numbers for *intI1* and 16S rRNA to be calculated from the Ct values. The number of eubacterial genomes was determined by dividing gene copy numbers by 5, the average number of 16S rRNA genes possessed by *Corynebacterium* and *Staphylococcus* species (http://rrndb.mmg.msu.edu/search.php) which are the most abundant members of the litter community.

## 2.7. Microbial Community Structure as Defined by Terminal Restriction Fragment Length Polymorphism (T-RFLP) of 16S rRNA PCR Amplicons

T-RFLP of community 16S rRNA was used to assess changes in microflora composition as previously described [45,50,51]. Each peak's position on a *Hae*III T-RFLP profile was matched against a Microsoft Access database of DNA fragment sizes predicted for 16S rDNA ribotypes [45] identified from a previously published litter metagene library [19], soil bacteria sequences available on the public databases [52,53] and litter sequences obtained by pyrosequencing in this study.

## 2.8. PCR Amplification of 16S rRNA for Pyrosequencing

PCR amplification of the V3 and V6 region of the bacterial 16S rRNA genes was conducted after an initial denaturation at 95 °C for 3 min followed by 20 cycles, of denaturation at 94 °C for 30 s, annealing at 60 °C for 30 s and extension at 68 °C for 60 s. The final extension was carried out at 68 °C for 4 min. The bacterial primers (*E. coli* numbers) were 515R-NK modified and the sevenfold-degenerate primer 27F YM + 3 [54,55,56]. For ease of handling bioinformatics of pyrosequencing, the primers were synthesized with a sequencing adaptor and a specific 8-nt barcode [57]. Primers were a gift from Dr. William Whitman (University of Georgia, Athens). DNA extracted from *Salmonella enterica* Typhimurium SR11 was used as control to test for errors during PCR amplification and pyrosequencing. PCR reactions were performed on an Idaho Rapid Cycler. PCR products were visualized by electrophoresis on 1% agarose gels, stained with SYBR Green Dye (Invitrogen, Carlsbad, CA, USA) and amplicons were excised from the gel. Amplicons obtained from 3 replicates of the same samples were pooled together. Products were purified from the agarose gel initially using the Qiagen QIAquick Gel Extraction Kit, followed by the Agencourt AMpure magnetic beads (Beckman Coulter, Brea, CA, USA). Purified DNA was resuspended in water, confirmed by gel electrophoresis then concentrations were determined using a spectrophotometer. Samples were submitted to the University of Georgia Genomics Facility for pyrosequencing according to methods established by the manufacturer.

## 2.9. Statistical and Sequence Analysis

The log *E. coli* counts and gene prevalence and abundance data were analyzed using SAS. T-RFLP datasets were analyzed for similarity using the correlation coefficient function of MS Excel applied to signal intensities of the fragment patterns occurring for each sample. Shannon diversity indices were calculated using the Merlin plugin for MS Excel [58]. Analysis of 16S rRNA sequences was carried out using MOTHUR v1.21.0 [59]. DNA from *Salmonella* Typhimurium was included as a quality control for the analysis following the procedure previously established [55]. Good quality sequences were aligned using the SILVA database in MOTHUR and further filtered. 16S sequences were loaded at MG-RAST [60] in order to generate tables of genus and species frequencies at 97% similarity. Distances were calculated on preclustered sequences, and operational taxonomic units (OTU) were formed using the average neighbor method in MOTHUR. Diversity indices and rarefaction curves were calculated in MOTHUR. Libshuff was used to estimate differences between libraries composition. Statistical analyses related to the frequency of specific sequences representing microorganisms present in samples were conducted by analysis of variance using the SAS software [61]. Significant differences were based on $p \leq 0.05$. Principal component analysis was performed using the vegan Community Ecology Package in R package version 2.0-7 (http://CRAN.R-project.org/package=vegan) [62]. To identify the prevalence of pathogenic bacteria, the FASTA sequence files deposited at MG-RAST (rast.nmpdr.org/) were compared to sequences deposited at the Ribosomal Database Project at 97% minimum similarity. The abundance of pathogens was estimated by quantifying the number of sequences exhibiting similarity to known pathogenic organisms. Statistical analysis related to the frequency of pathogens present in samples was conducted by analysis of variance using the SAS software.

# 3. RESULTS AND DISCUSSION

## 3.1. Effects of Antibiotic Usage Changes on Microbial Community Structure of Commercial Poultry Litter

In this study we removed antibiotic usage from two houses and changed the class of antibiotic that was used in two houses on three commercial farms; the study design is shown in Table 1, results for two farms are shown in Figure 1. Because of a freezer malfunction, many samples from the third farm were unavailable for testing. No treatment-associated changes were detected however inter-farm and flock differences were observed. In order to assess the composition of litter, the putative identity of each terminal fragment was inferred through *in silico* analysis of 16S rRNA sequences from a previous litter study [19] and soil bacterial sequences present in the public databases. Some species or strains of bacteria, such as *Aerococcus*, *Bacteroides*, and *Streptococcus* could not be distinguished by fragment size; therefore the compositional results are expressed as combinations of organisms. Overall results by farm and treatment are shown in Table 3 and Table 4; temporal results by farm, flock, and house are shown in Figure 1.

**Table 3** Diversity of bacterial communities determined by 16S rRNA T-RFLP signal of litter from 5 sequential flocks of broiler chickens raised on commercial poultry farms. Two houses on each farm contained birds that were fed the AGP flavomycin or flavomycin + ...

| | No antibiotic | | AGP | |
|---|---|---|---|---|
| | **Shannon Diversity Index** | **Richness** | **Shannon Diversity Index** | **Richness** |
| **Flock 1** | 3.94 ± 0.57 | 22.25 ± 4.6 | 4.11 ± 0.99 | 26.75 ± 13.1 |
| **Flock 2** | 3.68 ± 0.76 | 17.5 ± 8.3 | 3.76 ± 0.94 | 19.75 ± 13.3 |
| **Flock 3** | 3.85 ± 0.37 | 22.5 ± 6.4 | 3.66 ± 0.46 | 18.5 ± 3 |
| **Flock 4** | 3.42 ± 0.42 | 18 ± 5.6 | 3.38 ± 0.06 | 12.25 ± 8.73 |
| **Flock 5** | 3.27 ± 0.77 | 13.75 ± 6.2 | 3.83 ± 0.12 | 21.75 ± 6.8 |

**Table 4** Abundance of genera contributing greater than 2% of total 16S rRNA T-RFLP signal of litter from five sequential flocks of broiler chickens raised on commercial poultry farms. Two houses on each farm contained birds that were fed the AGP flavomycin or flavomycin + virginiamycin; two houses were fed no antibiotics.

| **Bacterial genus predicted by T-RFLP peak** | **No antibiotic *** | | **AGP *** | |
|---|---|---|---|---|
| | **Farm 1** | **Farm 2** | **Farm 1** | **Farm 2** |
| *Aerococcus, Bacteroides, Streptococcus* | 7.79% | 1.76% | 8.58% | 1.45% |
| *Clostridia, Ruminococcus, Eubacterium, Fecalobacterium* | 10.98% | 13.70% | 10.49% | 11.62% |
| *Corynebacterium, Lactobacillus* | 16.99% | 31.00% | 17.53% | 35.16% |
| *Corynebacterium, Brachybacterium, Joetgalibacillus/Salinicoccus* | 13.48% | 6.35% | 11.91% | 8.16% |
| *Lactobacillus* | 18.90% | 3.87% | 19.86% | 3.63% |
| *Staphylococcus* sp. | 7.20% | 4.75% | 7.26% | 3.62% |
| *Unknown* | 11.65% | 27.32% | 12.11% | 24.80% |
| *Yania, Clostridia* | 9.21% | 12.37% | 10.45% | 11.49% |

* No significant differences in composition were observed between treatments using Tukey test ($P > 0.05$).

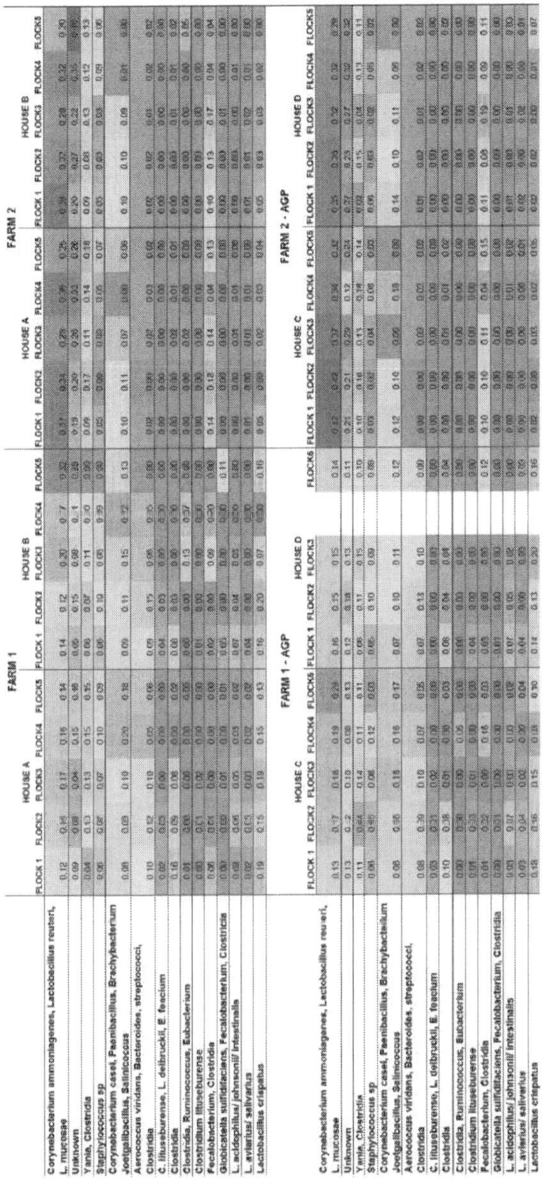

**Figure 1** Abundance of T-RFLP peaks contributing greater than 2% of total 16S rRNA signal of litter from 5 sequential flocks of broiler chickens reared on commercial poultry farms. Two houses (C and D) on each farm contained birds that were fed antibiotic growth promoters (AGP); houses A and B were fed no antibiotics. The results are shown as the proportion of total T-RFLP signal for each flock using a green → red gradient 0 0.05 0.1 0.15 0.2 0.25 0.3 0.35 0.4 0.45 0.5.

There were distinct differences in the litter composition between the two farms although no apparent differences were seen associated with antibiotic usage. Farm 1 tended to have more signal predicted to represent *Aerococcus/Bacteroides/Streptococcus*, corynebacteria, *Lactobacillus*, and *Staphylococcus* while farm 2 samples produced more signal for *Corynebacterium/Lactobacillus* and unknown bacteria (Table 4). Figure 1 suggests that the litter composition displayed community successions over time with flock cycles which partially resulted from litter management practices on the farms. Fresh pine shavings are commonly applied to the top layer of the litter before each flock is placed. However there appear to be differences in the composition of the litter from the first flock to the last and some differences between houses on the same farm.

In order to compare the composition of litter from different farms and treatments, Shannon diversity indices were used to detect whether there were differences in community structure based on the distribution of terminal fragments for each sample; correlation coefficients were used to determine similarity among the samples. Shannon diversity indices ranged from 3.27–4.11 with no significant differences noted by treatment or flock (Table 3). Richness ranged from 5–42 fragments with farm 1 mean richness (9 no-AGP, 13.75 AGP) similar to farm 2 (12.2 no-AGP, 12.5 AGP). Mean correlation coefficients were similar between houses despite treatment differences: farm 1 no-AGP 0.46, AGP 0.51; farm 2 no-AGP 0.88, AGP 0.84. When antibiotic-free houses were compared to AGP houses on the same farm, mean correlation coefficients ranged from 0.49–0.59 on farm 1 and 0.82–0.9 on farm 2. The lowest correlation coefficient (0.11) was associated with comparing the first flock cycle in one untreated house and one AGP house on farm 1, while the highest coefficient (0.98) was associated with comparing the two untreated houses for flock 1 on farm 2. In fact farm 2 correlation coefficients were consistently higher than those of farm 1 indicating significant differences among the study sites.

## 3.2. Effects of Antibiotic Usage Changes on the Prevalence of Antibiotic Resistance Genes within the Bacterial Community of Commercial Poultry Litter

Antibiotics were not used in two houses on each farm and two houses were changed from phosphoglycolipid to a streptogramin antibiotic in the finisher feed. PCR was used to detect the presence of streptogramin-resistance genes, *ermA*, *ermB*, *ermC*, *vatA*, *vatB*, *vatD*, *vatE*, and *vgaB*, tetracycline-resistance genes, *tetM*and *tetO*, and class 1 integon, *intI*1, within litter community DNA samples. The prevalence of the genes ranged from 0%–100% but no significant difference between treatments was detected (Table 5).

Class 1 integons, *ermA*, *ermB*, and *tetM* were detected in nearly every sample (83%–100% prevalence) while*tetO*, *vatA and vgaB* were primarily detected on farm 3. No *ermC* or *vatD* were detected in any sample.

## 3.3. Composition of Poultry Litter in Research Facilities Housing Birds Treated with Prebiotics or Probiotics

Because no significant changes in litter community structure were seen associated with five sequential commercial flocks grown without antibiotics, we sought to determine if litter could be changed by bacteriotherapy of the birds themselves. The flocks were administered commercial products containing bacteria or mannan oligosaccharides which are marketed to improve intestinal microbial ecology, reduce intestinal disease, and improve feed conversion and reduce *Salmonella* prevalence. We hypothesized that they may also affect the microbial ecology of litter therefore litter community DNA was isolated and subjected to pyrosequencing. After processing of 16S sequences for quality and length, 64 libraries produced 367,688 sequences with an average length of 316 bp. Individual library sizes ranged from 434 to 42,046 sequences. Seven samples failed to generate a significant number of reads, thus were removed from the study analysis. A total of 1,103 operational taxonomic units (OTUs) at the species level were obtained. The majority of OTUs contributed less than 10 total sequences however the genera that were detected were primarily members of the phyla *Firmicutes* and *Actinobacteria* (Table 6 and Figure 2) with *Firmicutes*contributing an average of 58% and *Actinobacteria* contributing an average of 28%. There were many genera in which only one or two sequences were detected which is a common finding [63]. While each treatment group had 9–15 different genera contributing a minimum of 1% of sequences to the total, in regards to the overall composition of samples, the Primalac and All-Lac + BioMos groups were the most diverse with greater than 150 genera detected. The control group was the least diverse with only 85 different genera detected.

The bacterial community of the first sample from the first cycle was statistically different ($P < 0.05$) from the last sample from the fourth cycle for all groups. The fact that the control house was statistically different indicates that time has a significant impact on the litter bacterial community. Table 7 shows the effect of time and treatments on litter composition. The first sample from all groups was dominated by the phylum Firmicutes however there was some variation in the most abundant genus. In the control group and All-Lac + BioMos treatment, *Staphylococcus* was most abundant while the other groups were dominated by*Salinococcus*. Firmicutes were also the most abundant phylum in the last sample in all treatment groups except the flocks receiving All-Lac + BioMos, which were dominated by the phylum Actinobacteria. Over time, only the Primalac group did not exhibit an increase in Actinobacteria. A decrease in the Firmicutes/Actinobacteria ratio from the first to last sample was observed in the control group compared to the average of the ratio observed in the litter from treated chickens (1.57 *versus* 1.87). This observation suggests that continued administration of bacteriotherapy may be necessary to maintain the abundance of the Firmicutes in the litter. *Corynebacterium*, *Brachybacterium* and *Brevibacterium* were the most abundant genera from the phylum Actinobacteria. This shift in bacterial communities over multiple production cycles is not unexpected given the physical differences between fresh and used litter. Past studies have shown that members of the phylum Actinobacteria may be involved in the decomposition of organic material such as the wood shavings used as bedding in this study [64].

**Table 5** Prevalence of resistance genes in litter obtained from 5 sequential flocks of broiler chickens raised on 3 commercial poultry farms. Two houses on each farm received no antibiotics; two houses received flavomycin or flavomycin + virginiamycin feed.

| TREATMENT * | | *ermA* | | | *ermB* | | | *ermC* | | | *tetM* | | | *tetO* | | | *vatA* | | | *vatB* | | | *vatD* | | *vatE* | | | *vgaB* | | | *intI1* | | |
|---|---|---|---|---|---|---|---|---|---|---|---|---|---|---|---|---|---|---|---|---|---|---|---|---|---|---|---|---|---|---|---|---|---|
| | | Farm | | | Farm | | | Farm | | | Farm | | | Farm | | | Farm | | | Farm | | | Farm | | Farm | | | Farm | | | Farm | | |
| | | 1 | 2 | 3 | 1 | 2 | 3 | 1 | 2 | 3 | 1 | 2 | 3 | 1 | 2 | 3 | 1 | 2 | 3 | 1 | 2 | 3 | 1 | 2 | 1 | 2 | 3 | 1 | 2 | 3 | 1 | 2 | 3 |
| **Flock 1** | Flavomycin | + | + | + | + | + | + | - | - | - | + | + | + | - | - | + | - | - | + | + | + | - | - | - | - | - | + | - | - | + | + | + | + |
| | | + | + | + | + | + | + | - | - | - | - | + | + | - | - | + | - | - | + | - | + | - | - | - | - | - | + | - | - | + | + | + | + |
| | NONE | + | + | + | + | + | + | - | - | - | + | + | + | - | - | + | - | - | + | + | + | - | - | - | - | + | + | - | - | + | + | + | + |
| | | + | + | + | + | + | + | - | - | - | + | + | + | - | - | + | - | - | + | - | + | - | - | - | - | + | + | - | - | + | + | + | + |
| **Flock 2** | Flavomycin + Virginiamycin | - | + | + | + | + | + | - | - | - | + | - | + | - | - | + | - | - | + | + | + | - | - | - | - | + | + | - | - | - | + | + | + |
| | | + | + | + | + | + | + | - | - | - | + | + | + | - | - | + | - | - | + | - | + | - | - | - | - | - | + | - | - | + | + | + | + |
| | NONE | + | + | + | + | + | + | - | - | - | + | + | + | - | - | + | - | - | + | + | + | - | - | - | - | + | + | - | - | + | + | + | + |
| | | + | + | + | + | + | + | - | - | - | + | + | + | - | - | + | - | - | + | - | + | - | - | - | - | - | + | - | - | + | + | + | + |
| **Flock 3** | Flavomycin + Virginiamycin | + | + | + | + | + | - | - | - | - | + | + | + | - | - | + | + | - | + | - | + | - | - | - | - | + | + | - | - | + | + | + | + |
| | | + | + | + | + | + | + | - | - | - | + | + | + | - | - | + | - | - | + | - | - | - | - | - | - | + | + | - | - | + | + | + | + |
| | NONE | + | + | + | + | + | - | - | - | - | + | - | + | - | - | + | - | - | + | - | - | - | - | - | - | - | + | - | - | + | + | + | + |
| | | + | + | + | + | + | + | - | - | - | + | - | + | - | - | + | - | - | + | - | - | - | - | - | - | - | + | - | - | + | + | + | + |
| **Flock 4** | Flavomycin + Virginiamycin | + | + | + | + | + | + | - | - | - | + | + | + | - | - | + | - | - | + | - | - | - | - | - | - | - | + | - | - | + | - | + | + |
| | | + | + | + | + | + | + | - | - | - | + | - | + | - | - | + | - | - | + | - | - | - | - | - | - | - | + | - | - | + | + | + | + |
| | NONE | + | - | + | + | + | + | - | - | - | - | - | + | - | - | + | - | - | + | - | - | - | - | - | - | + | + | - | - | + | + | + | + |
| | | - | - | + | + | + | + | - | - | - | + | — | + | - | - | + | - | - | + | - | + | - | - | - | - | - | + | - | - | + | + | + | + |
| **Flock 5** | Flavomycin | - | + | + | - | + | + | - | - | - | + | + | + | - | - | + | - | - | + | + | + | - | - | - | - | - | - | - | - | + | + | + | + |
| | | - | + | + | - | + | + | - | - | - | + | + | + | - | - | + | - | - | + | - | + | - | - | - | - | + | + | - | - | + | + | + | + |
| | NONE | - | + | + | + | + | + | - | - | - | + | + | + | - | - | + | - | - | + | - | + | - | - | - | - | + | + | - | - | + | + | + | + |
| | | + | - | + | + | + | + | - | - | - | + | + | + | - | - | + | - | - | + | - | + | - | - | - | - | + | - | - | - | + | + | + | + |
| **Total prevalence** | AGP | 90% | | | 90% | | | 0% | | | 90% | | | 33% | | | 33% | | | 33% | | | 0% | | 43% | | | 30% | | | 97% | | |
| | NONE | 83% | | | 97% | | | 0% | | | 87% | | | 33% | | | 33% | | | 30% | | | 0% | | 50% | | | 33% | | | 100% | | |

* No significant differences were observed among treatments using Cochran-Armitage test ($P > 0.05$).

**Table 6** Abundance and prevalence of genera comprising litter bacterial community of broiler chickens raised in research housing and administered probiotics or prebiotics. Number of 16S rRNA sequences exhibiting 97% similarity to each genus is shown with the percentage of sequences in parentheses.

| Genera | Control | All-Lac | All-Lac + BioMos | Aviguard | Primalac |
|---|---|---|---|---|---|
| Total sequences | 19,806 | 39,357 | 91,219 | 48,226 | 76,363 |
| Number of genera | 85 | 135 | 154 | 119 | 211 |
| *Staphylococcus* * | 5,605 (28.3) | 9,603 (24.4) | 32,171 (35.3) | 14,648 (30.4) | 19,138 (25.1) |
| *Corynebacterium* ** | 4,222 (21.3) | 9,483 (24.1) | 16,872 (18.5) | 7,839 (16.3) | 13,259 (17.4) |
| *Lactobacillus* * | 1,955 (9.9) | 3,925 (10.0) | 9,424 (10.3) | 4,767 (9.9) | 7,775 (10.2) |
| *Salinicoccus* * | 1,851 (9.3) | 3,412 (8.7) | 8,407 (9.2) | 5,802 (12.0) | 5,274 (6.9) |
| *Yaniella* * | 1,272 (6.4) | 1,796 (4.6) | 5,610 (6.2) | 2,467 (5.1) | 4,793 (6.3) |
| *Brachybacterium* ** | 1,077 (5.4) | 1,796 (4.6) | 3,534 (3.9) | 2,698 (5.6) | 3,066 (4.0) |
| *Brevibacterium* ** | 948 (4.8) | 1,371 (3.5) | 2,666 (2.9) | 2,758 (5.7) | 2,660 (3.5) |
| *Facklamia* * | 337 (1.7) | 1,291 (3.3) | 1,640 (1.8) | 796 (1.7) | 1,498 (2.0) |
| *Clostridium XI* * | 295 (1.5) | 424 (1.1) | 0 (0.0) | 556 (1.2) | 1,091 (1.4) |
| *Enterococcus* * | 0 (0.0) | 548 (1.4) | 0 (0.0) | 623 (1.3) | 974 (1.3) |
| *Atopostipes* * | 0 (0.0) | 0 (0.0) | 0 (0.0) | 0 (0.0) | 1,345 (1.8) |
| *Streptococcus* * | 0 (0.0) | 0 (0.0) | 0 (0.0) | 0 (0.0) | 1,260 (1.7) |

* Firmicutes; ** Actinobacteria.

| | flock | Age (weeks) | *Corynebacterium* | *Staphylococcus* | *Salinicoccus* | *Yaniella* | *Jeotgalicoccus* | *Brachybacterium* | *Brevibacterium* | *Lactobacillus* | *Enterococcus* | *Facklamia* | *Atopostipes* | *Clostridium_XI* |
|---|---|---|---|---|---|---|---|---|---|---|---|---|---|---|
| **Control** | 1 | 2 | 0.11 | 0.38 | 0.12 | 0.10 | 0.02 | 0.06 | 0.03 | 0.11 | 0.01 | 0.01 | 0.01 | 0.00 |
| | 1 | 5 | 0.24 | 0.24 | 0.10 | 0.06 | 0.05 | 0.10 | 0.06 | 0.04 | 0.03 | 0.03 | 0.00 | 0.01 |
| | 1 | 6 | 0.18 | 0.26 | 0.09 | 0.05 | 0.07 | 0.05 | 0.06 | 0.17 | 0.01 | 0.02 | 0.00 | 0.01 |
| | 2 | 2 | 0.21 | 0.33 | 0.09 | 0.05 | 0.05 | 0.11 | 0.07 | 0.02 | 0.01 | 0.01 | 0.00 | 0.01 |
| | 2 | 5 | 0.13 | 0.31 | 0.11 | 0.06 | 0.05 | 0.08 | 0.08 | 0.13 | 0.00 | 0.01 | 0.00 | 0.00 |
| | 3 | 1 | 0.35 | 0.22 | 0.07 | 0.05 | 0.08 | 0.05 | 0.05 | 0.04 | 0.01 | 0.03 | 0.01 | 0.01 |
| | 3 | 3 | 0.28 | 0.33 | 0.07 | 0.04 | 0.08 | 0.01 | 0.02 | 0.09 | 0.01 | 0.02 | 0.00 | 0.01 |
| | 3 | 6 | 0.16 | 0.41 | 0.06 | 0.06 | 0.04 | 0.02 | 0.04 | 0.09 | 0.02 | 0.01 | 0.00 | 0.06 |
| | 4 | 1 | 0.26 | 0.29 | 0.01 | 0.05 | 0.11 | 0.01 | 0.01 | 0.04 | 0.00 | 0.03 | 0.01 | 0.09 |
| | 4 | 3 | 0.25 | 0.17 | 0.12 | 0.08 | 0.09 | 0.07 | 0.05 | 0.07 | 0.00 | 0.02 | 0.00 | 0.02 |
| **All-Lac** | 1 | 1 | 0.13 | 0.03 | 0.24 | 0.05 | 0.02 | 0.13 | 0.00 | 0.00 | 0.00 | 0.05 | 0.01 | 0.00 |
| | 1 | 2 | 0.14 | 0.12 | 0.21 | 0.09 | 0.02 | 0.06 | 0.03 | 0.13 | 0.02 | 0.02 | 0.03 | 0.00 |
| | 1 | 5 | 0.24 | 0.19 | 0.10 | 0.06 | 0.06 | 0.09 | 0.05 | 0.04 | 0.03 | 0.09 | 0.02 | 0.01 |
| | 1 | 6 | 0.29 | 0.22 | 0.14 | 0.04 | 0.06 | 0.06 | 0.04 | 0.12 | 0.00 | 0.01 | 0.00 | 0.00 |
| | 2 | 2 | 0.19 | 0.12 | 0.12 | 0.04 | 0.08 | 0.04 | 0.05 | 0.26 | 0.04 | 0.01 | 0.00 | 0.00 |
| | 2 | 3 | 0.27 | 0.32 | 0.05 | 0.04 | 0.11 | 0.02 | 0.04 | 0.07 | 0.01 | 0.01 | 0.00 | 0.03 |
| | 2 | 5 | 0.23 | 0.48 | 0.03 | 0.02 | 0.08 | 0.01 | 0.02 | 0.06 | 0.01 | 0.02 | 0.00 | 0.01 |
| | 2 | 6 | 0.31 | 0.32 | 0.07 | 0.02 | 0.12 | 0.01 | 0.02 | 0.07 | 0.00 | 0.02 | 0.00 | 0.01 |
| | 3 | 1 | 0.34 | 0.00 | 0.06 | 0.02 | 0.09 | 0.02 | 0.03 | 0.06 | 0.00 | 0.04 | 0.01 | 0.01 |
| | 3 | 3 | 0.25 | 0.32 | 0.07 | 0.05 | 0.08 | 0.04 | 0.04 | 0.07 | 0.00 | 0.03 | 0.01 | 0.01 |
| | 3 | 6 | 0.24 | 0.30 | 0.04 | 0.04 | 0.08 | 0.04 | 0.03 | 0.13 | 0.00 | 0.02 | 0.00 | 0.03 |
| | 4 | 1 | 0.25 | 0.21 | 0.08 | 0.05 | 0.20 | 0.04 | 0.03 | 0.06 | 0.01 | 0.03 | 0.00 | 0.02 |
| | 4 | 3 | 0.24 | 0.22 | 0.08 | 0.06 | 0.08 | 0.01 | 0.04 | 0.17 | 0.00 | 0.01 | 0.00 | 0.03 |
| | 4 | 5 | 0.22 | 0.30 | 0.03 | 0.04 | 0.08 | 0.04 | 0.04 | 0.16 | 0.00 | 0.02 | 0.00 | 0.02 |
| **All-Lac + BioMos** | 1 | 1 | 0.14 | 0.31 | 0.16 | 0.09 | 0.10 | 0.04 | 0.03 | 0.06 | 0.01 | 0.02 | 0.00 | 0.02 |
| | 1 | 2 | 0.05 | 0.08 | 0.27 | 0.09 | 0.05 | 0.09 | 0.02 | 0.23 | 0.02 | 0.01 | 0.00 | 0.01 |
| | 1 | 5 | 0.19 | 0.45 | 0.07 | 0.05 | 0.06 | 0.03 | 0.03 | 0.09 | 0.01 | 0.01 | 0.00 | 0.00 |
| | 1 | 6 | 0.23 | 0.43 | 0.05 | 0.06 | 0.05 | 0.04 | 0.04 | 0.07 | 0.00 | 0.01 | 0.00 | 0.00 |
| | 2 | 2 | 0.26 | 0.19 | 0.07 | 0.05 | 0.08 | 0.07 | 0.03 | 0.13 | 0.02 | 0.04 | 0.00 | 0.00 |
| | 2 | 3 | 0.17 | 0.46 | 0.05 | 0.04 | 0.07 | 0.01 | 0.02 | 0.10 | 0.01 | 0.01 | 0.00 | 0.00 |
| | 2 | 5 | 0.20 | 0.38 | 0.06 | 0.05 | 0.08 | 0.04 | 0.04 | 0.09 | 0.00 | 0.02 | 0.00 | 0.01 |
| | 2 | 6 | 0.19 | 0.44 | 0.08 | 0.06 | 0.06 | 0.02 | 0.03 | 0.09 | 0.00 | 0.01 | 0.00 | 0.01 |
| | 3 | 1 | 0.30 | 0.27 | 0.09 | 0.07 | 0.08 | 0.04 | 0.04 | 0.04 | 0.01 | 0.03 | 0.00 | 0.01 |
| | 3 | 3 | 0.22 | 0.30 | 0.07 | 0.07 | 0.10 | 0.02 | 0.03 | 0.12 | 0.01 | 0.03 | 0.00 | 0.01 |
| | 4 | 3 | 0.13 | 0.26 | 0.09 | 0.07 | 0.12 | 0.02 | 0.02 | 0.11 | 0.03 | 0.01 | 0.00 | 0.04 |
| | 4 | 5 | 0.48 | 0.26 | 0.03 | 0.07 | 0.05 | 0.02 | 0.02 | 0.00 | 0.01 | 0.02 | 0.00 | 0.01 |
| **Aviguard** | 1 | 1 | 0.04 | 0.13 | 0.50 | 0.10 | 0.05 | 0.04 | 0.02 | 0.02 | 0.03 | 0.02 | 0.01 | 0.01 |
| | 1 | 2 | 0.05 | 0.41 | 0.12 | 0.07 | 0.01 | 0.07 | 0.05 | 0.11 | 0.03 | 0.01 | 0.00 | 0.00 |
| | 1 | 5 | 0.23 | 0.12 | 0.08 | 0.02 | 0.09 | 0.10 | 0.12 | 0.08 | 0.03 | 0.04 | 0.00 | 0.01 |
| | 1 | 6 | 0.13 | 0.38 | 0.08 | 0.03 | 0.06 | 0.08 | 0.07 | 0.12 | 0.00 | 0.01 | 0.00 | 0.00 |
| | 2 | 2 | 0.38 | 0.15 | 0.09 | 0.04 | 0.06 | 0.10 | 0.08 | 0.04 | 0.00 | 0.03 | 0.00 | 0.01 |
| | 2 | 3 | 0.12 | 0.37 | 0.07 | 0.03 | 0.04 | 0.04 | 0.07 | 0.22 | 0.01 | 0.01 | 0.00 | 0.00 |
| | 2 | 5 | 0.10 | 0.33 | 0.06 | 0.05 | 0.04 | 0.09 | 0.08 | 0.12 | 0.01 | 0.01 | 0.00 | 0.05 |
| | 2 | 6 | 0.14 | 0.41 | 0.06 | 0.04 | 0.06 | 0.06 | 0.08 | 0.08 | 0.00 | 0.01 | 0.00 | 0.01 |
| | 3 | 1 | 0.13 | 0.38 | 0.09 | 0.04 | 0.07 | 0.05 | 0.06 | 0.10 | 0.03 | 0.01 | 0.00 | 0.02 |
| | 3 | 3 | 0.14 | 0.41 | 0.07 | 0.04 | 0.06 | 0.04 | 0.10 | 0.05 | 0.03 | 0.01 | 0.00 | 0.01 |
| | 3 | 6 | 0.21 | 0.29 | 0.10 | 0.07 | 0.10 | 0.01 | 0.04 | 0.12 | 0.00 | 0.01 | 0.01 | 0.01 |
| | 4 | 1 | 0.29 | 0.22 | 0.06 | 0.05 | 0.19 | 0.03 | 0.03 | 0.05 | 0.00 | 0.02 | 0.00 | 0.02 |
| | 4 | 3 | 0.20 | 0.26 | 0.09 | 0.05 | 0.12 | 0.02 | 0.05 | 0.12 | 0.01 | 0.03 | 0.00 | 0.01 |
| | 4 | 5 | 0.30 | 0.20 | 0.08 | 0.05 | 0.11 | 0.02 | 0.03 | 0.00 | 0.04 | 0.12 | 0.01 | 0.01 |
| **Primalac** | 1 | 1 | 0.07 | 0.13 | 0.32 | 0.11 | 0.02 | 0.18 | 0.05 | 0.02 | 0.01 | 0.02 | 0.01 | 0.01 |
| | 1 | 5 | 0.20 | 0.19 | 0.12 | 0.07 | 0.03 | 0.06 | 0.06 | 0.06 | 0.01 | 0.03 | 0.03 | 0.01 |
| | 1 | 6 | 0.29 | 0.06 | 0.10 | 0.06 | 0.04 | 0.11 | 0.09 | 0.01 | 0.00 | 0.02 | 0.04 | 0.00 |
| | 2 | 2 | 0.08 | 0.08 | 0.03 | 0.02 | 0.03 | 0.02 | 0.01 | 0.48 | 0.20 | 0.01 | 0.00 | 0.01 |
| | 2 | 3 | 0.08 | 0.61 | 0.00 | 0.01 | 0.01 | 0.00 | 0.00 | 0.17 | 0.00 | 0.01 | 0.00 | 0.02 |
| | 2 | 5 | 0.27 | 0.43 | 0.01 | 0.03 | 0.11 | 0.02 | 0.01 | 0.06 | 0.00 | 0.01 | 0.01 | 0.00 |
| | 2 | 6 | 0.18 | 0.40 | 0.02 | 0.07 | 0.10 | 0.02 | 0.02 | 0.10 | 0.00 | 0.02 | 0.01 | 0.02 |
| | 3 | 1 | 0.16 | 0.31 | 0.01 | 0.06 | 0.09 | 0.01 | 0.02 | 0.24 | 0.03 | 0.01 | 0.02 | 0.01 |
| | 3 | 3 | 0.19 | 0.35 | 0.00 | 0.03 | 0.08 | 0.01 | 0.01 | 0.17 | 0.02 | 0.02 | 0.01 | 0.03 |
| | 3 | 6 | 0.18 | 0.41 | 0.01 | 0.10 | 0.07 | 0.02 | 0.03 | 0.10 | 0.00 | 0.01 | 0.00 | 0.02 |
| | 4 | 1 | 0.18 | 0.31 | 0.02 | 0.07 | 0.16 | 0.02 | 0.01 | 0.11 | 0.00 | 0.02 | 0.00 | 0.02 |
| | 4 | 3 | 0.25 | 0.30 | 0.01 | 0.06 | 0.07 | 0.02 | 0.01 | 0.09 | 0.02 | 0.01 | 0.00 | 0.05 |

**Figure 2** Abundance of genera contributing greater than 1% to total number of 16S rRNA sequences of 4 sequential flocks of broiler chickens administered probiotics or prebiotics and reared in research housing. The abundances are shown as the proportion of total sequences for each flock 

.

**Table 7** 16S rRNA composition of first and last litter sample from broiler chickens raised in research housing and administered probiotics or prebiotics.

| Treatment | Sample | Number of sequences | Number of genera | Shannon Diversity Index | *Firmicutes* (%) | *Actinobacteria* (%) | *Proteobacteria* (%) | Most abundant genus |
|---|---|---|---|---|---|---|---|---|
| Control | First | 3181 | 40 | 2.43 (±0.05) | 78.2 | 21.8 | 0.00 | *Staphylococcus* (37.6%) |
| | Last | 3068 | 43 | 2.72 (±0.05) | 61 | 38.9 | 0.00 | *Corynebacterium* (24.7%) |
| All-Lac | First | 2244 | 57 | 3.10 (±0.14) | 68.6 | 25.4 | 4.50 | *Salinicoccus* (24.4%) |
| | Last | 2888 | 25 | 2.36 (±0.05) | 68.7 | 30.8 | 0.00 | *Staphylococcus* (30.4%) |
| All-Lac + BioMos | First | 8949 | 72 | 2.59 (±0.03) | 78.1 | 21.8 | 0.05 | *Staphylococcus* (30.7%) |
| | Last | 587 | 24 | 1.89 (±0.13) | 47.4 | 52.6 | 0.00 | *Corynebacterium* (48%) |
| Aviguard | First | 4480 | 42 | 2.56 (±0.03) | 88.4 | 11.5 | 0.04 | *Salinicoccus* (49.9%) |
| | Last | 4468 | 42 | 2.40 (±0.03) | 64.1 | 35.1 | 0.72 | *Corynebacterium* (30%) |
| Primalac | First | 2938 | 37 | 2.75 (±0.05) | 66.4 | 32.5 | 0.27 | *Salinicoccus* (32%) |
| | Last | 2660 | 34 | 2.58 (±0.04) | 70.8 | 28 | 0.56 | *Staphylococcus* (30%) |

In order to examine the effect of the treatments on the litter composition, the sequences from the first sample from cycle one and the last sample from cycle 4 were compared to the corresponding samples in the control group. Treatment groups receiving All-Lac and Aviguard were found to be statistically different from the control group first sample ($p < 0.001$ and $p < 0.0001$, respectively). When examining the effect of treatment on the litter community from the last sample from cycle four, treatment groups receiving All-Lac + BioMos, Aviguard and Primalac were significantly different than the control group ($p < 0.0055$, $p < 0.0001$, $p < 0.0151$, respectively).

The Shannon diversity index was also used to compare composition of the litter community. The Shannon diversity index takes into consideration the different genera in a sample as well as the frequency of each species to quantify the entropy or uncertainty in a sample [65]. In other words, it quantifies the uncertainty of predicting a particular genus from a given sample, the greater the abundance, the more difficult it is to predict the next genera sampled. If there was only one genus present the

Shannon diversity index would be zero because there is no difficulty in predicting the correct genera. Using this ecological parameter, the control group increased in diversity from the first sample to the last sample while all of the treatment groups actually decreased in diversity (Table 7). A rarefaction curve comparing the occurrence of different sequences, also called operational taxonomic units (OTUs), demonstrated that diversity decreased over time (Figure 3(A)) probably due to changes in the physical/chemical characteristics of litter. Treatment comparisons using OTU rarefaction indicated richness was higher in litter from treated chickens and lower in the control group (Figure 3(B)). This indicates that the all of the treatments have the ability to impact composition of the litter bacterial community. Canonical Correspondence Analysis (Figure 4) shows that the litter community is strongly affected by its age. Most outliers observed were samples collected in the beginning of the experiment, which possessed a low abundance of *Actinobacteria* and low diversity, and litter from chickens receiving Primalac which had the highest diversity.

## 3.4. Effect of Time and Probiotic or Prebiotic Treatments on Pathogen Abundance in Litter

The litter DNA sequences were assessed for the presence of pathogenic species of bacteria and revealed that*Staphylococcus aureus* and *Clostridium perfringens* abundance was not affected by treatments or the cycle (data not shown). Abundance of *Clostridium sordellii* increased in litter over time but treatments did not affect abundance. Because *Salmonella* is known to contaminate poultry litter and it cannot be reliably detected using 16S rRNA, we screened multiple litter samples collected after the birds were 3 weeks of age from every flock by *Salmonella*-specific PCR in order to determine whether treatments decreased the prevalence of this pathogen (Table 8). *Salmonella* was detected within litter of every treatment group except flocks treated with Primalac. But, only the last flock of the All-Lac + BioMos group was *Salmonella*-positive while the other 3 groups had 50%–75% of flocks positive. These results indicate that bacteriotherapy can not only change the composition of poultry litter but can reduce the prevalence of pathogens.

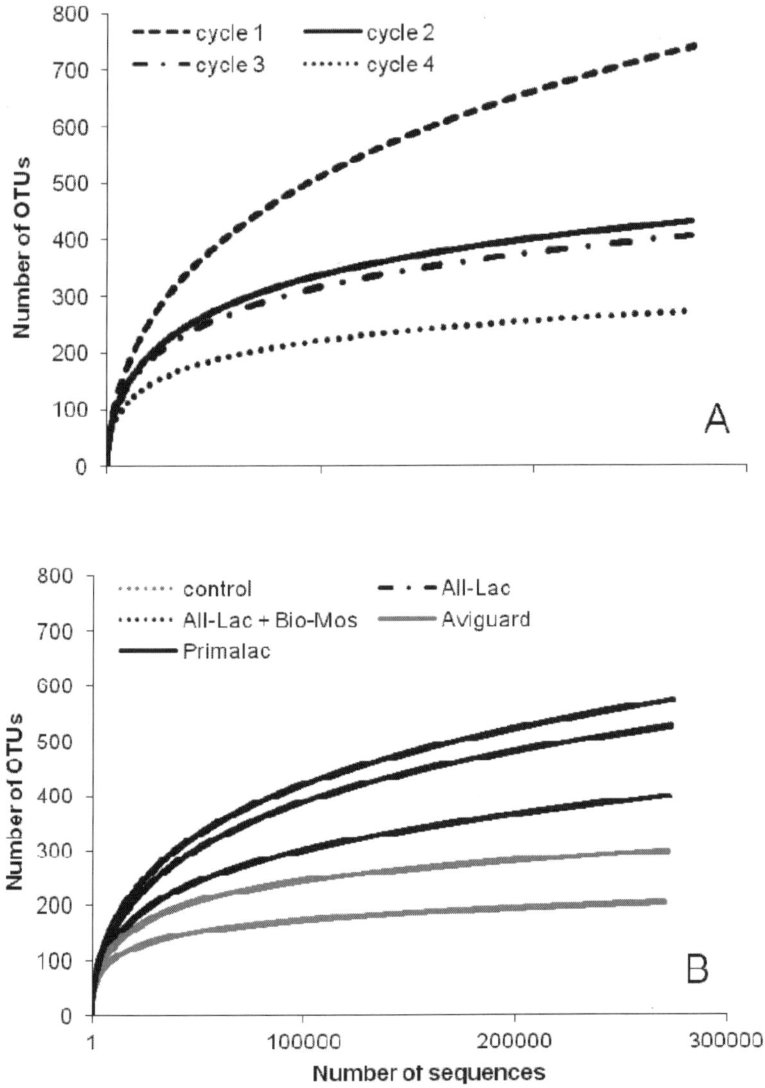


**Figure 3** Rarefaction curve for bacterial 16S rRNA gene OTUs at 97% of similarity for litter samples collected from 4 sequential flocks of broiler chickens administered probiotics or prebiotics and reared in research housing. Graph (**A**) analyzed by flock; graph (**B**) by treatment.

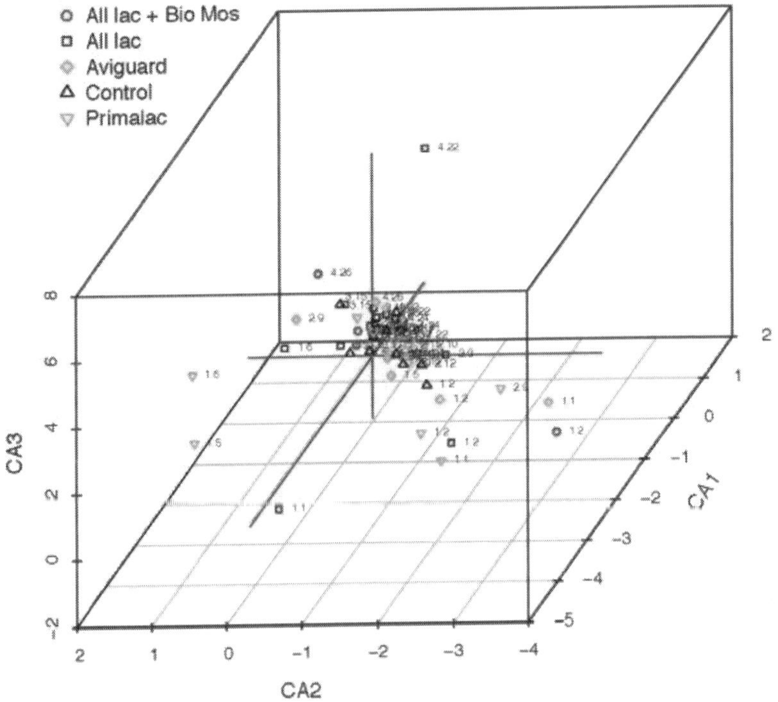


**Figure 4** Correspondence analysis of microbial community patterns generated by 16S rRNA analysis of litter samples collected from four sequential flocks of broiler chickens administered probiotics or prebiotics and reared in research housing. Each point represents a single sample; the five treatments are represented by color and symbol type as shown in the legend. Labels for each point indicate litter cycle (1–4), followed by week of sample collection (1–26).

**Table 8** Prevalence of *Salmonella* in litter of 4 sequential flocks of broiler chickens administered probiotics or prebiotics and reared in research housing.

| Flock | Control | All-Lac | All-Lac + BioMos | Aviguard | Primalac |
|---|---|---|---|---|---|
| 1 | + | - | - | + | - |
| 2 | + | + | - | - | - |
| 3 | - | + | - | + | - |
| 4 | - | + | + | + | - |

### 3.5. Effects of Probiotics and Prebiotics on the Prevalence of Antibiotic Resistance within the Bacterial Community of Poultry Litter

Two approaches were used to assay for changes in the litter resistome. Because tetracycline resistance has been shown to be high among poultry *E. coli* [66] and high levels of *tetM* were found in commercial poultry litter, we were interested in whether bacteriotherapy of birds would decrease the abundance of tetracycline-resistant coliforms. We quantified the abundance of resistant and susceptible coliforms for each litter sample by plating on selective agar containing antibiotic. Figure 5 shows the abundance of resistant coliforms nearly equals the abundance of susceptible isolates for each treatment and over every flock cycle. There were no statistically significant differences between treatments in total amounts of coliforms, amounts of tetracycline resistant coliforms, or the ratio of tetracycline resistant coliforms to total coliforms in the litter or the fecal samples (data not shown) for any of the treatments. Common belief is that competition resulting from bacteriotherapy would reduce the abundance of antibiotic-resistant organisms or genes because of the potential fitness cost of resistance [67]. However, it may be that the tetracycline resistance genes have no cost for these bacteria, or there has been compensatory adaptation to ameliorate the cost [67,68]. Therefore we assayed for the abundance of another common resistance element, the class 1 integron, because it has been found in a diversity of bacteria including the most abundant members of the litter community [16,69]. Integron abundance was determined by quantitative real-time PCR on litter community DNA and the signal normalized to eubacterial genome copies. The mean *intI1*/eubacterial genome ratio ranged from 0.028–0.047 suggesting that 3%–5% of the total litter community of each treatment group contained an integron (Table 9). There were no significant differences detected in abundance over time, flock or treatment suggesting that integron carriage is stable in poultry litter.

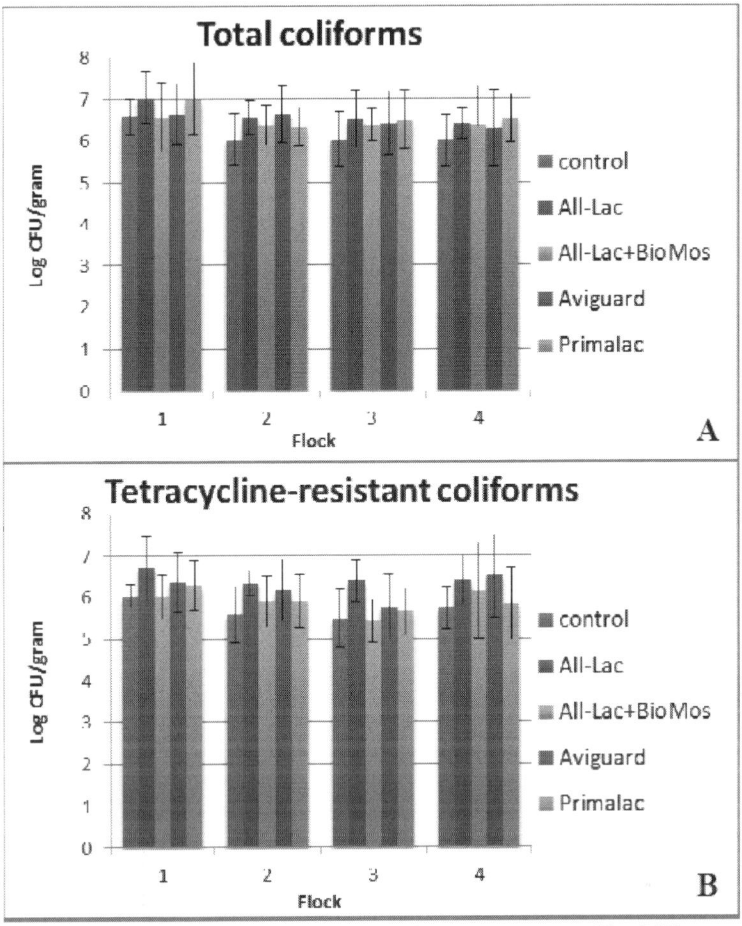


No significant differences were observed among treatments ($P > 0.05$).

**Figure 5** Log colony forming units (CFU) of coliforms (panel **A**) and tetracycline-resistant coliforms (panel **B**) cultured from litter collected from four sequential flocks of broiler chickens administered probiotics or prebiotics and reared in research housing.

**Table 9** Prevalence and abundance of class 1 integron (*intI1*) genes in litter obtained from four sequential flocks of broiler chickens administered probiotics or prebiotics and reared in research housing. The number (#) of eubacterial genomes and *intI1* copies per 25 ng litter DNA was calculated after normalizing the quantitative PCR signal using a control strain and adjusting the signal to reflect genome numbers for the most abundant genera in broiler litter.

| Treatment | # *intI1* copies * ± SD | # eubacterial genomes * ± SD | *intI1*: eubacterial genome ratio * | Mean Ratio ± SD |
|---|---|---|---|---|
| **Control** | | | | |
| Flock 1 | $6.30 \times 10^7 \pm 9.69 \times 10^6$ | $1.15 \times 10^9 \pm 1.23 \times 10^7$ | 0.0547 | |
| Flock 2 | $2.69 \times 10^7 \pm 2.90 \times 10^6$ | $1.26 \times 10^9 \pm 3.03 \times 10^8$ | 0.0214 | |
| Flock 3 | $3.11 \times 10^6 \pm 1.77 \times 10^5$ | $5.25 \times 10^8 \pm 1.05 \times 10^7$ | 0.0059 | |
| Flock 4 | $2.97 \times 10^7 \pm 8.46 \times 10^6$ | $1.04 \times 10^9 \pm 5.30 \times 10^7$ | 0.0286 | 0.0277 ± 0.0204 |
| **All-Lac** | | | | |
| Flock 1 | $3.30 \times 10^7 \pm 5.87 \times 10^5$ | $1.10 \times 10^9 \pm 2.54 \times 10^8$ | 0.0300 | |
| Flock 2 | $1.59 \times 10^7 \pm 9.31 \times 10^6$ | $5.23 \times 10^8 \pm 2.74 \times 10^7$ | 0.0304 | |
| Flock 3 | $3.15 \times 10^7 \pm 6.79 \times 10^6$ | $1.29 \times 10^9 \pm 2.25 \times 10^8$ | 0.0244 | |
| Flock 4 | $4.41 \times 10^7 \pm 7.29 \times 10^6$ | $2.07 \times 10^9 \pm 1.06 \times 10^9$ | 0.0213 | 0.0265 ± 0.0044 |
| **All-Lac + BioMos** | | | | |
| Flock 1 | $2.31 \times 10^7 \pm 4.12 \times 10^6$ | $7.26 \times 10^8 \pm 3.76 \times 10^7$ | 0.0318 | |
| Flock 2 | $3.38 \times 10^7 \pm 1.44 \times 10^6$ | $1.28 \times 10^9 \pm 1.29 \times 10^8$ | 0.0263 | |
| Flock 3 | $2.81 \times 10^7 \pm 2.09 \times 10^6$ | $1.10 \times 10^9 \pm 3.63 \times 10^7$ | 0.0254 | |
| Flock 4 | $2.37 \times 10^7 \pm 4.03 \times 10^6$ | $1.05 \times 10^9 \pm 4.58 \times 10^7$ | 0.0224 | 0.0265 ± 0.0039 |
| **Aviguard** | | | | |
| Flock 1 | $3.25 \times 10^7 \pm 1.06 \times 10^6$ | $1.14 \times 10^9 \pm 2.06 \times 10^8$ | 0.0286 | |
| Flock 2 | $3.56 \times 10^6 \pm 2.02 \times 10^5$ | $7.80 \times 10^7 \pm 4.63 \times 10^6$ | 0.0456 | |
| Flock 3 | $5.98 \times 10^7 \pm 3.42 \times 10^6$ | $8.17 \times 10^8 \pm 2.67 \times 10^7$ | 0.0732 | |
| Flock 4 | $4.18 \times 10^7 \pm 2.91 \times 10^6$ | $1.06 \times 10^9 \pm 2.16 \times 10^8$ | 0.0393 | 0.0467 ± 0.0190 |
| **Primalac** | | | | |
| Flock 1 | $5.69 \times 10^7 \pm 2.06 \times 10^7$ | $7.06 \times 10^8 \pm 6.80 \times 10^7$ | 0.0806 | |
| Flock 2 | $6.94 \times 10^7 \pm 2.61 \times 10^7$ | $1.33 \times 10^9 \pm 4.24 \times 10^5$ | 0.0520 | |
| Flock 3 | $2.76 \times 10^7 \pm 3.54 \times 10^5$ | $8.64 \times 10^8 \pm 6.04 \times 10^7$ | 0.0320 | |
| Flock 4 | $1.12 \times 10^7 \pm 7.07 \times 10^3$ | $4.27 \times 10^8 \pm 5.83 \times 10^7$ | 0.0263 | 0.0477 ± 0.0245 |

* No significant differences were observed among treatments using Tukey test ($P > 0.05$).

# 4. CONCLUSIONS

The deep litter system, where poultry litter is reused during consecutive cycles as a practice to reduce environmental waste, is common in many countries. However little is known about the impact of time or flock treatments on the microbial composition of poultry litter. Poultry litter under correct management is partially responsible for good animal production indices [70], because it acts as a source of beneficial organisms to colonize the intestinal tract of neonatal poultry which do not have exposure to the microbiota of adult birds [71]. Litter management is essential for disease control within the industry [72]. However, poultry litter has been shown to contain a diversity of pathogens, heavy metals, and have high levels of antibiotic resistance [14,16,19,73,74,75,76,77,78,79]. Studies investigating the stability of resistant organisms and resistance genes within litter have shown that some persist for long periods of time in stored litter suggesting that traditional storage methods may have little benefit [74,75,78]. In fact several studies have revealed increased prevalence and abundance of resistant organisms and resistance genes in soil that has been amended with litter or animal manure [11,15,18,80,81,82]. In this study we confirmed the high prevalence of resistance genes despite the absence of antibiotic selective pressure in multiple sequential flocks raised in the same poultry house. Using large scale libraries of litter 16S rRNA, from four cycles of antibiotic-free production with commercial alternatives to growth promoting antibiotics, we investigated whether these production changes would affect litter composition, abundance of antibiotic resistance and frequency of pathogens. Probiotics and prebiotics demonstrated the ability to alter the litter microbial community and some treatments reduced the prevalence of *Salmonella* in the litter. While the treatments changed the composition of the litter community, they did not reduce the prevalence or abundance of tetracycline-resistant *E. coli* or the class 1 integron resistance element, which is commonly detected in human and animal pathogens such as*Salmonella*, *E. coli*, *Shigella*, among others [69]. Application of litter and manures to land on which organic fruits and vegetables are produced could affect the prevalence of resistant bacteria on organic foods [83,84,85]. It is imperative that we identify methods to mitigate the abundance of resistance genes in litter and manures, however the full benefits of prebiotics and probiotics may require many, many cycles of production in order to affect the farm environment. Realizing environmental improvements from the usage of alternatives to antibiotics is in its infancy, as well as the study of the microbial ecology of broiler chicken production and its impact on the environment. Future research is needed to fully utilize these products in an effort to improve the environmental microbiome and resistome of food animal production.

## ACKNOWLEDGMENTS

This work was supported by the State of Georgia Veterinary Medical Experiment Station, grant 2009-03561 to MDL from the USDA National Institute of Food and Agriculture, grant 01-35212-10877 to CH from the USDA National Research Initiative and grant 99-35212-8680 to JJM from the USDA National Research Initiative.

## CONFLICTS OF INTEREST

The authors declare no conflict of interest. Coauthors Anne L. Hurley-Bacon, Andrea S. Zedek, and Gloria Avellaneda performed the research as part of the Masters of Avian Medicine or Ph.D. graduate programs prior to employment at their current affiliations.

## REFERENCES

1. US Department of Agriculture. Census of Agriculture; 2009. [(accessed on 25 March 2013)]. National Agricultural Statistics Service 2007. Available online: http://www.agcensus.usda.gov/Publications/2007/index.php.

2. Krider J. Agricultural Waste Management Field Handbook. US Department of Agriculture; Washington, DC, USA: 1992.

3. US Department of Agriculture. Agricultural Research Service. FY-2005 Annual Report Manure and Byproduct Utilization National

Program 206. 2008. [(accessed on 13 June 2013)]. Available online:http://www.ars.usda.gov/research/programs/programs.htm?np_code=206&docid=13337.

4. National Organic Program Rule-Soil Fertility and Crop Nutrient Management Practice Standard. [(accessed on 25 March 2013)]. Available online: http://www.ecfr.gov/cgi-bin/retrieveECFR?gp=&SID=2e169858257b8657bc0c02 3591b8183c&n=7y3.1.1.9.32.3&r=SUBPART&ty=HTML.

5. US Department of Agriculture. Economic Research Service. Organic Market Overview. 2012. [(accessed on 13 June 2013)]. Available online: http://www.ers.usda.gov/topics/natural-resources-environment/organic-agriculture/organic-market-overview.aspx#.UbnZfdhsOAI.

6. Jones P., Martin M. A Review of the Literature on the Occurrence and Survival of Pathogens of Animals and Humans in Green Compost. Waste and Resources Action Programme Standards Report. The Waste and Resources Action Programme; Banbury, UK: 2003.

7. Bicudo J., Goyal S. Pathogens and manure management systems: A review. Environ. Technol.2003;24:115–130. doi: 10.1080/ 0959333030 9385542.

8. King G., Brooks J.P., Brown S., Gerba C., O'Connor G.A., Pepper I.L. Land Application of Organic Residuals: Public Health Threat or Environmental Benefit? American Society for Microbiology; Washington, DC, USA: 2011.

9. Davis M.F., Price L.B., Liu C.M., Silbergeld E.K. An ecological perspective on U.S. industrial poultry production: The role of anthropogenic ecosystems on the emergence of drug-resistant bacteria from agricultural environments. Curr. Opin. Microbiol. 2011;14:244–250.

10. Gilchrist M.J., Greko C., Wallinga D.B., Beran G.W., Riley D.G., Thorne P.S. The potential role of concentrated animal feeding operations in infectious disease epidemics and antibiotic resistance. Environ. Health Perspect. 2007;115:313–316

11. Binh C.T., Heuer H., Kaupenjohann M., Smalla K. Diverse aadA gene cassettes on class 1 integrons introduced into soil via spread manure. Res. Microbiol. 2009;160:427–433.

12. Chee-Sanford J.C., Mackie R.I., Koike S., Krapac I.G., Lin Y.F., Yannarell A.C., Maxwell S., Aminov R. Fate and transport of antibiotic residues and antibiotic resistance genes following land application of manure waste. J. Environ. Qual. 2009;38:1086–1108

13. Davies J. Inactivation of antibiotics and the dissemination of resistance genes. Science. 1994;264:375–382.

14. Dhanarani T.S., Shankar C., Park J., Dexilin M., Kumar R.R., Thamaraiselvi K. Study on acquisition of bacterial antibiotic resistance determinants in poultry litter. Poult. Sci. 2009;88:1381–1387.

15. Heuer H., Schmitt H., Smalla K. Antibiotic resistance gene spread due to manure application on agricultural fields. Curr. Opin. Microbiol. 2011;14:236–243.

16. Nandi S., Maurer J.J., Hofacre C., Summers A.O. Gram-positive bacteria are a major reservoir of Class 1 antibiotic resistance integrons in poultry litter. Proc. Natl. Acad. Sci. USA. 2004;101:7118–7122.

17. Nwosu V.C. Antibiotic resistance with particular reference to soil microorganisms. Res. Microbiol.2001;152:421–430.

18. You Y., Hilpert M., Ward M.J. Detection of a common and persistent tet(L)-carrying plasmid in chicken-waste-impacted farm soil. Appl. Environ. Microbiol. 2012;78:3203–3213.

19. Lu J., Sanchez S., Hofacre C., Maurer J.J., Harmon B.G., Lee M.D. Evaluation of broiler litter with reference to the microbial composition as assessed by using 16S rRNA and functional gene markers. Appl. Environ. Microbiol. 2003;69:901–908.

20. Smith D.L., Harris A.D., Johnson J.A., Silbergeld E.K., Morris J.G., Jr. Animal antibiotic use has an early but important impact on the emergence of antibiotic resistance in human commensal bacteria. Proc. Natl. Acad. Sci. USA. 2002;99:6434–6439.

21. Gillings M.R., Stokes H.W. Are humans increasing bacterial evolvability? Trends Ecol. Evol.2012;27:346–352.

22. Schneitz C. Competitive exclusion in poultry—30 years of research. Food Control. 2005;16:657–667.

23. Kabir S.M. The role of probiotics in the poultry industry. Int. J. Mol. Sci. 2009;10:3531–3546.

24. Acar J., Casewell M., Freeman J., Friis C., Goossens H. Avoparcin and virginiamycin as animal growth promoters: A plea for science in decision-making. Clin. Microbiol. Infect. 2000;6:477–482.

25. Schwarz S., Chaslus-Dancla E. Use of antimicrobials in veterinary medicine and mechanisms of resistance. Vet. Res. 2001;32:201–225.

26. Gunal M., Yayli G., Kaya O., Karahan N., Sulak O. The effects of antibiotic growth promoter, probiotic or organic acid supplementation on performance, intestinal microflora and tissue of broilers. Int. J. Poult. Sci.2006;5:149–155.

27. Gustafson R.H., Bowen R.E. Antibiotic use in animal agriculture. J. Appl. Microbiol. 1997;83:531–541.

28. Fairchild A.S., Grimes J.L., Jones F.T., Wineland M.J., Edens F.W., Sefton A.E. Effects of hen age, Bio-Mos, and Flavomycin on poult susceptibility to oral *Escherichia coli* challenge. Poult. Sci. 2001; 80: 562–571.

29. Nurmi E., Rantala M. New aspects of Salmonella infection in broiler production. Nature. 1973;241:210–211.

30. Hooge D.M. Meta-analysis of broiler chicken pen trials evaluating dietary mannan oligosaccharide. Int. J. Poult. Sci. 2004;3:163–174.

31. Waldroup P.W., Oviedo-Rondon E.O., Fritts C.A. Comparison of Bio-Mos and antibiotic feeding programs in broiler diets containing copper sulfate. Int. J. Poult. Sci. 2003;2:28–31.

32. Fuller R. Probiotics in man and animals. J. Appl. Bacteriol. 1989;66:365–378.

33. Nava G.M., Bielke L.R., Callaway T.R., Castañeda M.P. Probiotic alternatives to reduce gastrointestinal infections: The poultry experience. Anim. Health Res. Rev. 2005;6:105–118.

34. Griggs J.P., Jacob J.P. Alternatives to antibiotics for organic poultry production. J. Appl. Poult. Res.2005;14:750–756.

35. Al-Zenki S.F., Al-Nasser A.Y., Al-Saffar A.E., Abdullah F.K., Al-Bahouh M.E., Al-Haddad A.S., Alomirah H., Mashaly M. Effects of using a chicken-origin competitive exclusion culture and probiotic cultures on reducing Salmonella in broilers. J. Appl. Poult. Res. 2009;18:23–29.

36. Carina Audisio M., Oliver G., Apella M.C. Protective effect of Enterococcus faecium J96, a potential probiotic strain, on chicks infected with Salmonella Pullorum. J. Food Prot. 2000;63:1333–1337.

37. Carter A.J., Adams M.R., Woodward M.J., La Ragione R.M. Control strategies for Salmonella colonization of poultry: The probiotic perspective. Food Sci. Technol. 2009;5:103–115.

38. Cox J.M., Chung B.L. Competitive exclusion: Probiotic preparations for poultry. Proc. Aust. Poult. Sci. Symp. 2000;12:32–39.

39. Crippen T.L., Sheffield C.L., Andrews K., Dowd S.E., Bongaerts R.J., Nisbet D.J. Planktonic and biofilm community characterization and Salmonella resistance of 14-day-old chicken cecal Microflora-derived continuous-flow cultures. J. Food Prot. 2008;71:1981–1987.

40. Feng Y., Gong J., Yu H., Jin Y., Zhu J., Han Y. Identification of changes in the composition of ileal bacterial microbiota of broiler chickens infected with Clostridium perfringens. Vet. Microbiol. 2010;140:116–121.

41. Hajati H., Hassanabadi A. The effects of probiotic on growth performance and mortality of broiler chickens; Proceedings of the 3rd International Veterinary Poultry Congress; Tehran, Iran. 22–23 February 2012; p. 117.

42. Hammons S., Oh P.L., Martínez I., Clark K., Schlegel V.L., Sitorius E., Scheideler S.E., Walter J. A small variation in diet influences the

Lactobacillus strain composition in the crop of broiler chickens. Syst. Appl. Microbiol. 2010;33:275–281.

43. Hofacre C.L., Froyman R., Gautrias B., George B., Goodwin M.A., Brown J. Use of Aviguard and other intestinal bioproducts in experimental Clostridium perfringens-associated necrotizing enteritis in broiler chickens. Avian Dis. 1998;42:579–584.

44. Hofacre C.L., Primm N.D., Vance K., Goodwin M.A., Brown J. Comparison of a lyophilized chicken-origin competitive exclusion culture, a lyophilized probiotic, and fresh turkey cecal material against Salmonella colonization. J. Appl. Poult. Res. 2000;9:195–203.

45. Lu J., Hofacre C., Smith F., Lee M.D. Effects of feed additives on the development on the ileal bacterial community of the broiler chicken. Animal. 2008;2:669–676.

46. Rantala M., Nurmi E. Prevention of the growth of Salmonella infantis in chicks by the flora of the alimentary tract of chickens. Br. Poult. Sci. 1973;14:627–630.

47. Apajalahti J.H., Särkilahti L.K., Mäki B.R., Heikkinen J.P., Nurminen P.H., Holben W.E. Effective recovery of bacterial DNA and percent-guanine-plus-cytosine-based analysis of community structure in the gastrointestinal tract of broiler chickens. Appl. Environ. Microbiol. 1998;64:4084–4088.

48. Simjee S., McDermott P.F., White D.G., Hofacre C., Berghaus R.D., Carter P.J., Stewart L., Liu T., Maier M., Maurer J.J. Antimicrobial susceptibility and distribution of antimicrobial-resistance genes among Enterococcus and coagulase-negative Staphylococcus isolates recovered from poultry litter. Avian Dis.2007;51:884–892.

49. Liu T., Liljebjelke K., Bartlett E., Hofacre C., Sanchez S., Maurer J.J. Application of nested polymerase chain reaction to detection of Salmonella in poultry environment. J. Food Prot. 2002;65:1227–1232.

50. Buchan A., Newell S.Y., Butler M., Biers E.J., Hollibaugh J.T., Moran M.A. Dynamics of bacterial and fungal communities on decaying salt marsh grass. Appl. Environ. Microbiol. 2003;69:6676–6687.

51. Lu J., Hofacre C.L., Lee M.D. Emerging technologies in microbial ecology aid in understanding the effect of monensin on necrotic enteritis. J. Appl. Poult. Res. 2006;15:145–153.

52. SILVA: Comprehensive Ribosomal RNA Databases. 2012. [(accessed on 11 November 2012)]. Available online: http://www.arb-silva.de/search/

53. Genomic BLAST. [(accessed on 11 November 2012)]. Available online:http://www.ncbi.nlm.nih.gov/sutils/genom_table.cgi.

54. Acosta-Martinez V., Dowd S., Sun Y., Allen V. Tag encoded pyrosequencing analysis of bacterial diversity in a single soil type as affacted by management and land use. Soil Biol. Biochem. 2008;40:2762–2770.

55. Garcia S.L., Jangid K., Whitman W.B., Das K.C. Transition of microbial communities during the adaption to anaerobic digestion of carrot waste. Bioresour. Technol. 2011;102:7249–7256.

56. Frank J.A., Reich C.I., Sharma S., Weisbaum J.S., Wilson B.A., Olsen G.J. Critical evaluation of two primers commonly used for amplification of bacterial 16S rRNA genes. Appl. Environ. Microbiol.2008;74:2461–2470.

57. Hamady M., Walker J.J., Harris J.K., Gold N.J., Knight R. Error-correcting barcoded primers for pyrosequencing hundreds of samples in multiplex. Nat. Methods. 2008;5:235–237.

58. Millar N. Merlin Version 3.0. [(accessed on 12 March 2012)]. Available online:http://www.heckgrammar.co.uk/index.php?p=10310.

59. Schloss P.D., Westcott S.L., Ryabin T., Hall J.R., Hartmann M., Hollister E.B., Lesniewski R.A., Oakley B.B., Parks D.H., Robinson C.J., et al. Introducing mothur: Open-source, platform-independent, community-supported software for describing and comparing microbial communities. Appl. Environ. Microbiol. 2009;75:7537–7541.

60. Meyer F., Paarmann D., D'Souza M., Olson R., Glass E.M., Kubal M., Paczian T., Rodriguez A., Stevens R., Wilke A., et al. The metagenomics RAST server-a public resource for the automatic

phylogenetic and functional analysis of metagenomes. BMC Bioinform. 2008;9:386.

61. SAS-Procedures Guide and Statistical Procedures. SAS Institute Inc.; Cary, NC, USA: 2008.

62. Oksanen J., Blanchet F.G., Kindt R., Legendre P., Minchin P.R., O'Hara R.B., Simpson G.L., Solymos P., Stevens M.H.H., Wagner H. vegan: Community Ecology Package. R package version 2.0-7. 2004. [(accessed on 14 December 2012)]. Available online: http://CRAN.R-project.org/package=vegan.

63. Dumas M.D., Polson S.W., Ritter D., Ravel J., Gelb J., Jr., Morgan R., Wommack K.E. Impacts of poultry house environment on poultry litter bacterial community composition. PLoS One. 2011;6

64. Yang Y., Iji P.A., Kocher A., Mikkelsen L.L., Choct M. Effects of mannanoligosaccharide on growth performance, the development of gut microflora, and gut function of broiler chickens raised on new litter. J. Appl. Poult. Res. 2007;16:280–288.

65. Hill T.C., Walsh K.A., Harris J.A., Moffett B.F. Using ecological diversity measures with bacterial communities. FEMS Microbiol. Ecol. 2003;43:1–11.

66. Smith J.L., Drum D.J., Dai Y., Kim J.M., Sanchez S., Maurer J.J., Hofacre C.L., Lee M.D. Impact of antimicrobial usage on antimicrobial resistance in commensal Escherichia coli strains colonizing broiler chickens. Appl. Environ. Microbiol. 2007;73:1404–1414.

67. Andersson D.I., Levin B.R. The biological cost of antibiotic resistance. Curr. Opin. Microbiol.1999;2:489–493. doi: 10.1016/S1369-5274(99)00005-3.

68. Maisnier-Patin S., Berg O.G., Liljas L., Andersson D.I. Compensatory adaptation to the deleterious effect of antibiotic resistance in Salmonella typhimurium. Mol. Microbiol. 2002;46:355–366.

69. Rowe-Magnus D.A., Mazel D. The role of integrons in antibiotic resistance gene capture. Int. J. Med. Microbiol. 2002;292:115–125.

70. Vieira S., Moran E. Effects of delayed placement and used litter on broiler yields. J. Appl. Poult. Res.1999;8:75–81.

71. Cressman M.D., Yu Z., Nelson M.C., Moeller S.J., Lilburn M.S., Zerby H.N. Interrelations between the microbiotas in the litter and in the intestines of commercial broiler chickens. Appl. Environ. Microbiol.2010;76:6572–6582.

72. Fussell L.W. Poultry industry strategies for control of immunosuppressive diseases. Poult. Sci.1998;77:1193–1196.

73. Sims J.T., Wolf D.C. Poultry Waste Management: Agricultural and Environmental Issues. In: Donald L.S., editor. Advances in Agronomy. Academic Press; New York, NY, USA: 1994. pp. 1–83.

74. Kelley T.R., Pancorbo O.C., Merka W.C., Thompson S.A., Cabrera M.L., Barnhart H.M. Fate of selected bacterial pathogens and indicators in fractionated poultry litter during storage. J. Appl. Poult. Res.1994;3:279–288.

75. Kelley T.R., Pancorbo O.C., Merka W.C., Thompson S.A., Cabrera M.L., Barnhart H.M. Bacterial pathogens and indicators in poultry litter during re-utilization. J. Appl. Poult. Res. 1995;4:366–373.

76. Kelley T.R., Pancorbo O.C., Merka W.C., Barnhart H.M. Antibiotic resistance of bacterial litter isolates.Poult. Sci. 1998;77:243–247.

77. Khan A., Nawaz M.S., Khan S.A., Steele R. Detection and characterization of erythromycin-resistant methylase genes in Gram-positive bacteria isolated from poultry litter. Appl. Microbiol. Biotechnol.2002;59:377–381.

78. Graham J.P., Evans S.L., Price L.B., Silbergeld E.K. Fate of antimicrobial-resistant enterococci and staphylococci and resistance determinants in stored poultry litter. Environ. Res. 2009;109:682–689.

79. Graham J.P., Price L.B., Evans S.L., Graczyk T.K., Silbergeld E.K. Antibiotic resistant enterococci and staphylococci isolated from flies collected near confined poultry feeding operations. Sci. Total Environ.2009;407:2701–2710.

80. Zhai Q., Coyne M., Barnhisel R. Mortality rates of fecal bacteria in subsoil amended with poultry manure. Bioresour. Technol. 1995;54:165–169.

81. Sistani K., Bolster C.H., Way T.R., Tobert H.A., Pote D.H., Watts D.B. Influence of poultry litter application methods on the longevity of nutrient and *E. coli* in runoff from tall fescue pasture. Water Air Soil Pollut. 2010;206:3–12.

82. Byrne-Bailey K.G., Gaze W.H., Zhang L., Kay P., Boxall A., Hawkey P.M., Wellington E.M.H. Integron prevalence and diversity in manured soil. Appl. Environ. Microbiol. 2011;77:684–687.

83. Wright G.D. Antibiotic resistance in the environment: A link to the clinic? Curr. Opin. Microbiol.2010;13:589–594.

84. Ruimy R., Brisabois A., Bernede C., Skurnik D., Barnat S., Arlet G., Momcilovic S., Elbaz S., Moury F., Vibet M.A., et al. Organic and conventional fruits and vegetables contain equivalent counts of Gram-negative bacteria expressing resistance to antibacterial agents. Environ. Microbiol. 2010;12:608–615.

85. Duffy E., Lucia L.M., Kells J.M., Castillo A., Pillai S.D., Acuff G.R. Concentrations of *Escherichia coli*and genetic diversity and antibiotic resistance profiling of Salmonella isolated from irrigation water, packing shed equipment, and fresh produce in Texas. J. Food Prot. 2005;68:70–79.

# CHAPTER 7

# Isolation and characterization of H9N2 influenza virus isolates from poultry respiratory disease outbreak

**Subhash J Jakhesara[1], Vaibhav D Bhatt[1], Namrata V Patel[1], Kantilal S Prajapati[2] and Chaitanya G Joshi[1]**

1Department of Animal Biotechnology, College of Veterinary Science & Animal Husbandry, Anand Agricultural University, Anand, Gujarat 388001, India.

2Department of Veterinary Pathology, College of Veterinary Science & Animal Husbandry, Anand Agricultural University, Anand, Gujarat 388001, India.

## ABSTRACT

The present study reports isolation and characterization of H9N2 virus responsible for disease characterized by symptoms including difficulty in respiration, head swelling, nasal discharge, reduced feed intake, cyanotic comb, reduced egg production and mortality. Virus isolation from allantoic fluid inoculated with tracheal aspirates and whole genome sequencing of two isolates were performed on an Ion-Torrent sequencer. Phylogenetic analysis revealed that the two H9N2 isolates are reassortant viruses showing a G1-like lineage for HA, NA and NP, a Hok/49/98-like lineage for PB1 and PA, PK/UDL-01/05-like lineage for PB2, IL/90658/00-like lineage for NS and an unknown lineage for M gene. Analyses of the HA cleavage site showed a sequence of (333PARSSR↓GL340) indicating that these isolates are of low pathogenicity. Isolate 2 has leucine at amino acid position 226, a substitution which is associated with mammalian adaptation of avian influenza virus. Isolate 1 has the S31N substitution in the M2 gene that has been associated with drug resistance as well as R57Q and C241Y mutations in the NP gene which are associated with human adaptation. The result reported here gives deep insight in to H9N2 viruses circulating in domestic poultry of India and supports the policy of active efforts to control and manage H9N2 infections in Indian poultry.

## KEYWORDS

Avian influenza, Phylogeny, Poultry, Next generation sequencing, Virus isolation

## 1. INTRODUCTION

Influenza viruses are negative strand RNA viruses of the Orthomyxoviridae family that cause a variety of infections of high and low pathogenicity in hosts that include mammalian and avian species. Influenza viruses are divided in to types A, B and C. Type B and Type C cause mild infection only in human being and show limited genetic diversity as they are restricted to humans. In contrast type A viruses co-circulate in different host species and display considerable genetic diversity that is acquired through genetic reassortments and mutations. Over time a variety of influenza A subtypes have adapted to poultry hosts from wild birds and these can pose serious zoonotic risks to humans. Since the initiation of influenza surveillance in 2001 India has observed infections from various subtypes including H11N1 (Pawar et al. 2010), H4N6 (Pawar et al. 2012), H5N1 (Mishra et al. 2009), H9N2 (Tosh et al. 2008) in avian; H1N1, H3N2 (Chadha et al. 2012) H5N1

(ABY27653, unpublished) in humans and H3N8 (Virmani et al. 2010) in horse. H9N2 is enzootic in birds in India and causes infection with low pathogenicity.

H9N2 infection in domestic poultry was reported during 2003–04 in poultry farms with history of reduced egg production and respiratory illness in Punjab, Haryana, Uttar Pradesh, Gujarat and Orissa states of India (Nagarajan et al. 2009). Similar cases of respiratory illness were reported from layer poultry farms in the Sarsa village of Gujarat, India. Clinical disease signs, such as difficulty in breathing, head swelling, nasal discharge, reduced feed intake and cyanotic comb, were recorded and affected flocks showed illness for 3–4 weeks with a drop in egg production of up to 40%. Approximately, 2–3% and 10–30% of mortality was observed in affected grower and layer flocks, respectively. Postmortem examination revealed fibrino-caseative material in the infra-orbital sinus, congested tracheal mucosa with caseous material, oviduct wall edema, fibrin plaques on the ovary and caseous plugs in the tracheal bifurcation.

The genetic, antigenic and evolutionary characterization of H9N2 viruses circulating in India is not well documented. Here, to explore the genetic and phylogenetic nature of H9N2 viruses in India, we have isolated H9N2 viruses from clinically affected birds and used whole genome sequencing to analyze two independent isolates at the molecular level. In particular we have examined these isolates for molecular signatures indicative of potential adaptation of viruses to mammalian host species.

# 2. MATERIALS AND METHODS

## 2.1 Virus isolation

Tracheal aspirates and swabs were collected from dead chickens during pathological examination. The collected clinical specimens were centrifuged for 10 minutes at 8, 000 × g, and after removing bacterial contaminants by millipore membrane filters (0.22 μm), the supernatant was incubated in the allantoic cavities of 10-day-old specific pathogen free (SPF) chicken eggs at 37°C for 72 h for viral isolation; finally the allantoic fluids were harvested and stored at −80°C.

## 2.2 Viral RNA extraction and sequencing

Isolation of viral RNA was done from virus containing allantoic fluid using a QIAmp viral RNA mini kit (Qiagen, Germany). The quality, quantity and integrity of total RNA was evaluated using a NanoDrop1000 spectrophotometer (Thermo Fisher Scientific, DE) and a Bioanalyzer 2100 (Agilent Technologies, CA). Total RNA was subjected to emulsion PCR after library preparation and DNA-positive beads from emulsion PCR were recovered, enriched and subjected to Ion semiconductor sequencing using a 316 chip on an Ion-Torrent sequencer following manufacturer's protocol.

## 2.3 Quality filtering and mapping of reads

Available reads were subjected to quality filtering by PrinseQ (Schmieder and Edwards 2011) based on the criteria of a minimum read length 40 and mean base quality for the entire read of 25. Quality filtered reads were assembled against the genome sequence of H9N2 strain A/Quail/Hong_Kong/G1/97 using CLC genomics workbench v4.7.1 (http://www.clcbio.com/products/clc-genomics-workbench/).

## 2.4 Genetic and phylogenetic analysis

Sequence analysis was conducted using Mega 5.2 (Tamura et al. 2011). Nucleotide sequence alignments were constructed for each of the eight genomic segments using the MUSCLE algorithm in Mega 5.2. The H9N2 isolates in the current study, the H9N2 virus strains isolated from India, all available Asian strains with full genome sequences from avian and humans and three different H9N2 virus lineages specified by the National Center for Biotechnology Information (NCBI) influenza database were included in the analysis. All sequence data were downloaded from the NCBI influenza database. The neighbour-joining method with Kimura two-parameter distances was used to construct the phylogenetic trees using Mega 5.2 (Kimura 1980). The reliability of internal branches was assessed by the p-distance substitution model using 1000 bootstrap replications.

## 2.5 Deduced amino acid sequence analysis

Consensus sequences of each segment were subjected to blastx against the NCBI flu database. Regions of amino acid sequences showing maximum match against flu protein sequences were considered for further analysis. Amino acid sequences from each segment were subjected to multiple alignments using MUSCLE. To determine variations at the sites of viral attachment, we used a web based application, ScanProsite (de Castro et al. 2006) to compare and identify N-glycosylation sites and determined differences with other strains.

# 3. RESULTS

## 3.1 Sequencing of influenza virus isolates

A total of 14,914 and 314,197 reads were obtained respectively for isolate1 and isolate 2. After stringent quality filtering and duplicate removing 4,457 (29.88%) and 137,675 (43.82%) sequences were assembled against genome sequence of H9N2 strain A/Quail/Hong Kong/G1/97. Alignments were manually observed for homopolymer and indel verification and curated consensus sequences were produced for each segment of both the isolates.

## 3.2 Phylogenetic analysis

Eight gene segments of the 2 isolates were phylogenetically analyzed against full length sequences of 199 Asian viruses as well as all available Indian H9N2 viral sequences (Additional file 1). Lineages for each of segment were assigned according to the previous study on comprehensive phylogenetic analysis of H9N2 viruses isolated from 1966 to 2009 (Dong et al. 2011). The phylogenetic tree of the HA gene (segment 4) revealed that both isolates were grouped together with other Indian strains (Additional file 1: Figure S1A). The phylogenetic tree of HA genes revealed that the G1 like lineage (represented by A/Quail/Hong Kong/G1/97) is the lineage under which all the Indian viruses can be grouped. A Similar profile was obtained for the NA gene (segment 6) and all Indian strains together with two isolates can be grouped under G1 like lineage (Additional file 1: Figure S1B). Similar to HA and NA, the phylogenetic tree of the NP gene (segment 5) revealed that all Indian strains could be grouped under G1 like lineage (Additional file 1: Figure S1F). The PB2 gene (segment 1) of Pakistan/UDL-01/2005 like lineage was predominant in the Indian strains (Additional file 1: Figure S1C). Phylogenetic tree for PB1 (segment 2; Additional file 1: Figure S1D) and PA (segment 3; Additional file 1: Figure S1E) genes revealed that all the Indian H9N2 strains together with two isolates from this study could be grouped under the Hok/49/98 like lineage for both the genes (represented by A/duck/Hokkaido/49/98). For NS gene (segment 8; Additional file 1: Figure S1H), Indian strains together with two isolates clustered under IL/90658/00 like lineage (represented by A/ck/Israel/90658/2000). Interestingly for MP gene (segment 7; Additional file 1: Figure S1G) no known lineage can be assigned and all Indian strains examined, along with H9N2 strains from Pakistan and Israel, formed a distinct cluster.

## 3.3 Molecular analysis of amino acid sequence

The deduced amino acid sequences of all genes were compared with representative Indian strains containing different HA and NA serotypes. It has been shown that the presence of multiple basic amino acids at the cleavage position increases cleavage of HA in to HA1 and HA2 and may result in increased pathogenicity (Kawaoka and Webster 1988). Deduced amino acid sequences at the cleavage site of both the isolates lacked multiple basic amino acids; comparison to other Indian H9N2 strains revealed a similar RSSR motif with exception of some strains in which the fourth amino acid upstream from the cleavage site was mutated to Lysine (K)/Valine (V) (Table 1). It needs to be studied whether this change in cleavage site has any effect on pathogenicity. It is considered that glycosylation sites affect the receptor binding capacity of HA (Ohuchi et al. 1997). Analysis of HA proteins of G1 lineage strain has revealed 8 potential N-glycosylation sites at 29, 105, 141, 206, 218, 298, 305, and 492 whilst the isolates identified in this study lack two potential glycosylation sites present at 206 and 218. It has been suggested that glycosylation patterns plays an important role in the adaptation of avian influenza virus to poultry (Matrosovich et al. 1999) hence it is possible that these changes in sites may contribute to selective avian adaptation.

**Table 1** Comparison of amino acid residues at specific positions in the HA and NA gene

| Strain | HA | | | | | NA | | | | | |
|---|---|---|---|---|---|---|---|---|---|---|---|
| | Receptor binding site | | | | Cleavage site | Stalk deletion | | HB site | | | |
| | 189 | 190 | 226 | 227 | 333–340 | 38–39 | 62–65 | 366–373 | 402 | 403 | 432 |
| Isolate1 | T | A | L | I | PARSSR↓GL | No | No | IKKDSRAG | N | W | R |
| Isolate2 | T | A | Q | I | PARSSR↓GL | No | No | IKKDSRAG | N | W | Q |
| A/chicken/Chandigarh/2048/2003 | T | A | L | I | PARSSR↓GL | No | No | IKKDSRAG | N | W | NA |
| A/chicken/Haryana/2424/2004 | T | A | L | I | PARSSR↓GL | No | No | IKKDSRAG | N | W | NA |
| A/chicken/India/IVRI-0C04/2010 | T | A | L | I | PAKSSR↓GL | No | No | IKRDSRAG | N | W | Q |
| A/chicken/India/WB-NIV1057231/2010 | T | V | L | I | PARSSR↓GL | No | No | IKEDSRAG | N | W | Q |
| A/chicken/Orissa/2317/2004 | T | A | L | I | PARSSR↓GL | No | No | IKKDSRAG | N | W | NA |
| A/chicken/Tripura/105131/2008 | T | A | L | I | PAKSSR↓GL | No | No | INKDSRAG | N | W | R |
| A/chicken/Uchal/8293/2006 | T | A | L | I | PARSSR↓GL | No | No | IRKDSREG | N | L | Q |
| A/watercoot/Haryana/5844/2005 | T | A | L | I | PARSSR↓GL | No | No | IKKDSRAG | N | W | Q |
| A/Quail/Hong Kong/G1/97 H9N2 | T | E | L | Q | PARSSR↓GL | Yes | No | IKKDSRSG | I | R | Q |
| A/duck/Hong Kong/Y280/97H9N2 | T | T | L | Q | PARSSR↓GL | No | Yes | IKEDSRSG | N | W | Q |
| A/turkey/Wisconsin/1/1966H9N2 | T | E | Q | Q | PAVSSR↓GL | No | No | ISKDSRSG | N | W | Q |

Receptor binding site in HA protein was also analyzed with respect to amino acid changes at positions 189,190,226 and 227 (H3 numbering, 197,198,234 and 235 for H9 numbering). Both the current isolates together with all Indian H9N2 strains (excluding A/ck/India/WB-NIV1057231/2010-Valine) have alanine at position 190, compared to the three characterized H9N2 lineage strains (G1, Y280, WI1) which contain glutamic acid or threonine. Additionally, all the Indian strains (excluding A/duck/India/31g/86-Q) and both the current isolates have isoleucine (I) at position 227 (H3 numbering, 235 for H9 numbering) in contrast to glutamine (Q) in all three lineage strains (Table 1). Most importantly the substitution Q226L was observed in isolate 2 in present study, which is associated with mammalian adaptation of avian influenza virus.

The deduced amino acid sequence of the NA protein was compared with that of Indian isolates and lineage defining strains. None of the current isolates or any of the other Indian strains possesses NA stalk deletions at positions 38–39 and 63–65 in the G1 and Y280 lineages respectively. Sequence analysis of the sialic acid-binding haemabsorbing (HB) site located on the surface of the NA molecule at positions 366–373, 400, 403 and 432 (Kobasa et al. 2001) is well conserved with some exceptions. A common substitution resulting in change of serine to alanine or glutamic acid was observed at position 372 amongst the isolates and Indian strains, respectively. Isolate1 along with only one Indian strain (A/ck/Tripura/105131/08) also revealed a unique substitution from glutamine to arginine at position 432 when compared with strains of G1 and Y280 lineage. Comparison of conserved potential N-glycosylation sites in the NA glycoprotein showed that the lineage strains contain seven sites at positions 61, 69, 86, 146, 200, 234 and 402. The isolates sequenced in this study as well as most of the Indian isolates revealed an additional site at asparagine (N) 44 due to substitution of P45S.

Several amino acid changes present in other proteins of influenza virus are known to play roles as molecular determinants for host range and virulence. Several such determinants have been described in various studies (Finkelstein et al. 2007; Shaw et al. 2002; Chen et al. 2006; Manzoor et al. 2009; Naffakh et al. 2008; Katz et al. 2000). The isolates sequenced in this study were also analyzed for these signatures in different genes (PB1, PB2, PA, NP, MA and M2) and results are described in Table 2. Substitutions of 2, 1, 2, 3, 1, 3 and 1 residues from avian associated H9 viruses to human associated H9 viruses (or other amino acids) were found in genes PB2, PB1, PA, NP, M1, M2 and NS1 respectively in current isolates. NS2 protein did not revealed any substitution frequently observed in human associated influenza viruses.

**Table 2** Analysis of mutation at amino acid positions important for host range pathogenicity and virulence

| Protein | aa position | Deduced amino acid | | Amino acid in isolates from this study | Reference |
|---|---|---|---|---|---|
| | | Human | Avian | | |
| PB2 | 44 | S | A | A | (Chen et al. 2006; Finkelstein et al. 2007; Shaw et al. 2002) |
| | 64 | T | M | M | (Finkelstein et al. 2007) |
| | 81 | M | T | T | (Shaw et al. 2002) |
| | 199 | S | A | A | (Chen et al. 2006; Shaw et al. 2002; Finkelstein et al. 2007) |
| | 256 | G | D | D | (Manzoor et al. 2009) |
| | 271 | A | T | T | (Chen et al. 2006; Shaw et al. 2002) |
| | 333 | I | T | T | (Naffakh et al. 2008) |
| | 355 | Q | K | **R**[a] | (Katz et al. 2000) |
| | 475 | M | L | L | (Chen et al. 2006) |
| | 482 | R | K | K | (Naffakh et al. 2008) |
| | 588 | I | A | **V** | (Chen et al. 2006; Shaw et al. 2002; Finkelstein et al. 2007) |
| | 613 | T | V | V | (Chen et al. 2006; Shaw et al. 2002) |
| | 627 | K | E | E | (Chen et al. 2006; Naffakh et al. 2008) |
| | 661 | T | A | A | (Shaw et al. 2002) |
| | 674 | T | A | A | (Chen et al. 2006; Shaw et al. 2002; Finkelstein et al. 2007) |
| | 701 | N | D | D | (Naffakh et al. 2008) |
| | 702 | R | K | K | (Shaw et al. 2002) |
| | 674 | T | A | A | (Chen et al. 2006; Shaw et al. 2002; Finkelstein et al. 2007) |
| | 701 | N | D | D | (Naffakh et al. 2008) |
| | 702 | R | K | K | (Shaw et al. 2002) |
| | 714 | R | S | S | (Naffakh et al. 2008) |
| PB1 | 13 | P | L | **P** | (Naffakh et al. 2008) |
| | 327 | K | R | R | (Chen et al. 2006) |
| | 336 | I | V | V | (Chen et al. 2006) |
| | 538 | G | D | D | (Naffakh et al. 2008) |
| | 578 | Q | K | K | (Naffakh et al. 2008) |
| | 678 | N | S | S | (Naffakh et al. 2008) |
| PA | 28 | L | P | P | (Chen et al. 2006; Shaw et al. 2002) |
| | 55 | N | D | D | (Chen et al. 2006; Shaw et al. 2002) |
| | 57 | Q | R | **Q**/R | (Chen et al. 2006; Finkelstein et al. 2007) |
| | 65 | L/Y | S | S* | (Naffakh et al. 2008; Shaw et al. 2002) |
| | 100 | A | V | V | (Shaw et al. 2002) |
| | 133 | G | E | E | (Naffakh et al. 2008) |
| | 241 | Y | C | **Y**/C | (Shaw et al. 2002) |
| | 225 | C | S | S | (Chen et al. 2006) |
| | 268 | I | L | L | (Chen et al. 2006) |
| NP | 16 | D | G | G | (Chen et al. 2006) |
| | 31 | K | R | R | (Shaw et al. 2002) |

| | | | | | |
|---|---|---|---|---|---|
| | 33 | I | V | V | (Chen et al. 2006; Shaw et al. 2002) |
| | 34 | N | D | **G** | (Naffakh et al. 2008) |
| | 61 | L | I | I | (Chen et al. 2006; Shaw et al. 2002) |
| | 100 | V | R | R | (Chen et al. 2006; Shaw et al. 2002) |
| | 109 | V | I | I | (Chen et al. 2006) |
| | 127 | D | E | E | (Shaw et al. 2002) |
| | 136 | M | L | **I** | (Shaw et al. 2002) |
| | 214 | K | R | R | (Chen et al. 2006; Shaw et al. 2002) |
| | 283 | P | L | L | (Chen et al. 2006; Shaw et al. 2002) |
| | 319 | K | N | N | (Naffakh et al. 2008) |
| | 293 | K | R | R | (Chen et al. 2006; Shaw et al. 2002) |
| | 305 | K | R | R | (Chen et al. 2006) |
| | 313 | Y | F | F | (Chen et al. 2006; Shaw et al. 2002) |
| | 357 | K | Q | Q | (Chen et al. 2006; Finkelstein et al. 2007) |
| | 372 | D | E | **D** | (Chen et al. 2006) |
| | 375 | G/E | D | D | (Shaw et al. 2002) |
| | 422 | K | R | R | (Chen et al. 2006) |
| | 442 | A | T | T | (Chen et al. 2006) |
| | 455 | E | D | D | (Chen et al. 2006) |
| | 480 | N | D | D | (Naffakh et al. 2008) |
| M1 | 15 | I | V | **I** | (Katz et al. 2000) |
| | 115 | I | V | V | (Chen et al. 2006) |
| | 121 | A | T | T | (Chen et al. 2006) |
| | 137 | A | T | T | (Shaw et al. 2002) |
| M2 | 11 | I | T | T | (Chen et al. 2006) |
| | 16 | G/D | E | **G** | (Shaw et al. 2002) |
| | 20 | N | S | S | (Chen et al. 2006; Shaw et al. 2002) |
| | 28 | I/V | I | **F** | (Shaw et al. 2002) |
| | 55 | F | L | **F** | (Shaw et al. 2002) |
| | 57 | H | Y | Y | (Chen et al. 2006) |
| | 78 | K | Q | Q | (Shaw et al. 2002) |
| | 86 | A | V | V | (Chen et al. 2006) |
| NS1 | 215 | T | P | P | (Finkelstein et al. 2007) |
| | 227 | R/K | E | **K** | (Chen et al. 2006; Finkelstein et al. 2007) |
| NS2 | 70 | G | S | S | (Chen et al. 2006) |

[a]Amino acids in boldface denotes the change from nucleotide commonly observed in avian associated H9 viruses.

# 4. DISCUSSION

Avian influenza is a serious threat to the poultry industry's economy and continuous surveillance and vaccination are two main factors in control of disease. Since 2006, India has experienced three outbreaks that caused severe economic losses (Chakrabarti et al. 2009). As a response to these outbreaks, surveillance of avian influenza in India has become a priority and several reports of avian influenza surveillance have been published (Pawar et al. 2012; Rao 1987). Avian influenza surveillance has confirmed the H9N2 strain as predominant along with H4N6, causing low pathogenic avian influenza in wet poultry markets and backyard poultry in West Bengal. Because influenza A viruses continue to evolve, the identification and characterization of local strains circulating at regional level is of paramount importance to keep a constant watch on influenza enzootics. The genetic nature of H9N2 viruses showing potential human receptor binding specificity and increasing reports of H9N2 viruses in India, along with newly emerging reassortant viruses, puts H9N2 in the list of viruses posing potential threat to humans.

This study represents a comprehensive, in-depth analysis of H9N2 viruses isolated from Gujarat, India. We could show that both isolates in this study are reassortant in nature; i.e. they contain more than one lineage for the eight segments present in their genome. Lineage assignment revealed that H9N2 viruses circulating in Gujarat comprise genetically similar segments from viral strains of other Asian countries such as China, Japan, Israel and Pakistan. This may be attributed to typical routes taken by migratory wild birds migrating from east-Asia through India, giving opportunities for viruses to exchange segments during co-circulation (Pawar et al. 2012). Phylogenetic analysis revealed an unknown lineage for the MP gene which was further separated in to two sub lineages consisting of strains reported before and after 2008 with the exception of two strains A/ck/India/IVRI-004/2010 and A/ck/India/IVRI-0011/2011 which were grouped with strains isolated before 2008. To check the homology of the MP gene of our isolates to strains other than H9N2 we performed blast and blastx of segment 7 nucleotide sequence. We found 96% and 99% of homology to a H7N3 strain from Pakistan (A/chicken/Karachi/NARC-100/2004) at nucleotide and protein level, respectively. Indeed, it has been found from several studies of H9N2 strains from Pakistan (Iqbal et al. 2009) and China (Guan et al. 2000), that reassortant H9N2 strain contains segments from highly pathogenic H5N1 or H7N3 strains. As LPAI H9N2 is widely prevalent in domestic poultry, it is possible that novel reassortant virus with high pathogenicity may arise and the virus may infect humans. This warrants influenza surveillance of migratory birds coming to India for validation.

The host range and adaptation to different host is correlated with amino acids present within and around the HA receptor binding site (RBS) which can confer different receptor binding specificity (Weis et al. 1988). In particular the amino acids glutamine (Q) or leucine (L) at position 226 (H3 numbering, 234H9 numbering) plays a key role in avian and human virus-like receptor specificity respectively. The presence of leucine (L) at position 226 is associated with mammalian adaptation avian influenza virus. In our study except for position 226, no additional changes were observed in the HA RBS that could enable avian-human transmission. Similar type of observations were also reported for HA gene of H9N2 viruses prevalent in Israel after detailed molecular analysis and it was found that H9N2 viruses circulating in the Israel are predominantly of G1 lineage and harbors Q226L substitution in HA gene (Davidson et al. 20132014). The Q226L substitution seen in RBS of H9N2 avian isolates is important as it confers human virus like receptor specificity and binds more strongly with 2-6-linked sialic acid receptors as seen in human H3N2 viruses (Matrosovich et al. 2001) in contrast to viruses lacking this substitution which binds more strongly with 2-3 linked sialic acid receptors. To check when this substitution was first introduced in Indian isolates we compared amino acid sequences of the HA gene and we found that H9N2 virus was isolated for the first time in India from the duck in 1986 which did not contain this substitution. Thereafter again it was detected in 2003 from Chandigarh and Haryana state of India and we found that these viruses which introduced after 2003 contained Q226L substitution. Our analysis found that since 2003 Q226L substitution is prevailing in the H9N2 viruses predominantly of G1 like lineage isolated from the India. This report shows need of continuous surveillance and reporting of the molecular changes occurring in the avian H9N2 viruses which can confer human virus like receptor specificity.

Taking account of amino acid changes in other internal genes of the isolates sequenced in this study, 2, 1, 2, 3, 1, 3 and 1 substitutions were found in the PB2, PB1, PA, NP, M1, M2 and NS1 genes. The PB2 gene in both isolates revealed K355R and A588V substitutions which are not associated with human associated H9 viruses. Additionally, both viruses have a L13P substitution frequently observed in human associated H9 viruses, but it is important to note that this change is commonly observed in H9N2 strains. The NP gene revealed D34G, L136I and E372D substitutions frequently seen in human associated H9 viruses. Considering the PA gene, isolate1 revealed R57Q and C241Y substitutions, which are associated with human adaptation. However, isolate 2 has retained avian associated amino acids at these positions. Whether these changes in PB2, PB1, PA and NP genes will cause increased pathogenicity and virulence in chickens or have emerged as a result of adaptation to hosts other than avian remains to be investigated, nonetheless substitutions at these positions are considered to influence replication efficiency and may act as hallmarks of host-specific adaptation. The M1 protein revealed V15I substitution in both isolates which is common in all the H9N2 lineages. The M2 gene has been extensively studied for drug resistance several mutations are known to confer drug resistance; L26F, V27A, A30T, S31N, G34E, L38F (Schnell and Chou 2008). Isolate 1 from this study has the S31N substitution, which has also been identified in H9N2 viruses from Iran, UAE and Qatar (Fusaro et al. 2011). Analyses of additional host range signatures for M2 gene in both isolates revealed I28F, E16G and L55F substitutions commonly observed in human associated H9 viruses. These changes in M1 and M2 proteins are likely to confer phenotypes of increased replication efficiency and amantadine resistance. Our analysis of the NS1 protein revealed E227K substitution in both the isolates, a substitution that confers increased pathogenicity and virulence in mice (Jackson et al. 2008). This lies within the C-terminal 227KSEV230 PDZ binding motif that is implicated in virus inhibition of apoptosis and overcoming host interferon responses (Thomas et al. 2011). E227K substitution is observed in some of the H5N1 viruses isolated from Indonesia and Saudi Arabia (Obenauer et al. 2006; Monne et al. 2008). However, it remains to be investigated whether E227K substitution results in increased virulence of H9N2 viruses in avian species. It is important to note here that we assessed these substitutions probably associated with host range and pathogenicity for H9N2 viruses identified in present study only, whether these substitutions are prevailing in other zoonotic influenza subtypes and contributes towards variation in pathogenicity needs further study and validation.

## 5. CONCLUSION

This study represents comprehensive genetic and phylogenetic characterization of H9N2 viruses recently isolated from chickens in Gujarat, India. Phylogenetic analyses suggest that there is continuous evolution of H9N2 viruses with development of reassortant virus showing intra and inter subtype exchange of segments. A probable role for migratory birds in development of reassortant viruses could be validated by surveillance of avian influenza viruses in migratory birds and large scale sequencing efforts to characterize H9N2 in India. Further, the strains identified in this study harbor some substitutions which are frequently observed in human associated H9 viruses which need to be further studied by animal models for their effect on pathogenicity. The results reported herein gives deep insight in to H9N2 viruses circulating in domestic poultry of India, however there is a need for further characterization by sequencing a greater number of isolates. This study supports policies for the active control and management of H9N2 infections in Indian poultry markets to reduce the risk of transmission to human.

## ACKNOWLEDGEMENTS

Authors are very thankful to Professor Fiona Tomley, Chair of Experimental Parasitology, Royal Veterinary College, UK for necessary corrections and help in improvement of this manuscript. Authors are also thankful to Professor M. K. Jhala, Veterinary Microbiogy, AAU, Anand for improvement of the manuscript.

## COMPETING INTEREST

None of the authors of this paper has a financial or personal relationship with other people or organizations that could inappropriately influence or bias the content of the paper.

## AUTHORS' CONTRIBUTIONS

KSP conducted sample collection and virus isolation. SJJ and VDB performed sequencing of the isolates. NP helped SJJ in bioinformatics analysis of data. SJJ analyzed and interpreted the data and wrote the manuscript. CGJ coordinated and helped to draft the manuscript. All authors read and approved the final manuscript.

# REFERENCES


1 Chadha MS, Broor S, Gunasekaran P, Potdar VA, Krishnan A, Chawla-Sarkar M, Biswas D, Abraham AM, Jalgaonkar SV, Kaur H, Klimov A, Lal RB, Moen A, Kant L, Mishra AC: Multisite virological influenza surveillance in India: 2004-2008. *Influenza Other Respir Viruses* 2012, 6(3):196-203. doi:10.1111/j.1750-2659.2011.00293.x 10.1111/j.1750-2659.2011.00293.x

2 Chakrabarti AK, Pawar SD, Cherian SS, Koratkar SS, Jadhav SM, Pal B, Raut S, Thite V, Kode SS, Keng SS, Payyapilly BJ, Mullick J, Mishra AC: Characterization of the influenza A H5N1 viruses of the 2008-09 outbreaks in India reveals a third introduction and possible endemicity. *PloS One* 2009, 4(11):e7846. doi:10.1371/journal. pone. 0007846 10.1371/journal.pone.0007846

3 Chen GW, Chang SC, Mok CK, Lo YL, Kung YN, Huang JH, Shih YH, Wang JY, Chiang C, Chen CJ, Shih SR: Genomic signatures of human versus avian influenza A viruses. *Emerg Infect Dis* 2006, 12(9):1353-1360.

4 Davidson I, Shkoda I, Golender N, Perk S, Lapin K, Khinich Y, Panshin A: Genetic characterization of HA gene of low pathogenic H9N2 influenza viruses isolated in Israel during 2006-2012 periods. *Virus Genes* 2013, 46(2):255-263.

5 Davidson I, Fusaro A, Heidari A, Monne I, Cattoli G: Molecular evolution of H9N2 avian influenza viruses in Israel. *Virus Genes* 2014. doi:10.1007/s11262-014-1037-0

6 de Castro E, Sigrist CJ, Gattiker A, Bulliard V, Langendijk-Genevaux PS, Gasteiger E, Bairoch A, Hulo N: ScanProsite: detection of PROSITE signature matches and ProRule-associated functional and structural residues in proteins. *Nucleic Acids Res* 2006, 34(Web Server issue):W362-W365.

7 Dong G, Luo J, Zhang H, Wang C, Duan M, Deliberto TJ, Nolte DL, Ji G, He H: Phylogenetic diversity and genotypical complexity of H9N2 influenza A viruses revealed by genomic sequence analysis. *PloS One* 2011, 6(2):e17212.

8 Finkelstein DB, Mukatira S, Mehta PK, Obenauer JC, Su X, Webster RG, Naeve CW: Persistent host markers in pandemic and H5N1 influenza viruses. *J Virol* 2007, 81(19):10292-10299.

9 Fusaro A, Monne I, Salviato A, Valastro V, Schivo A, Amarin NM, Gonzalez C, Ismail MM, Al-Ankari AR, Al-Blowi MH, Khan OA, Maken Ali AS, Hedayati A, Garcia Garcia J, Ziay GM, Shoushtari A, Al Qahtani KN, Capua I, Holmes EC, Cattoli G: Phylogeography and evolutionary history of reassortant H9N2 viruses with potential human health implications. *J Virol* 2011, 85(16):8413-8421.

10 Guan Y, Shortridge KF, Krauss S, Chin PS, Dyrting KC, Ellis TM, Webster RG, Peiris M: H9N2 Influenza Viruses Possessing H5N1-Like Internal Genomes Continue To Circulate in Poultry in Southeastern China. *J Virol* 2000, 74(20):9372-9380.

11 Iqbal M, Yaqub T, Reddy K, McCauley JW: Novel genotypes of H9N2 influenza A viruses isolated from poultry in Pakistan containing NS genes similar to highly pathogenic H7N3 and H5N1 viruses. *PloS One* 2009, 4(6):e5788.

12 Jackson D, Hossain MJ, Hickman D, Perez DR, Lamb RA: A new influenza virus virulence determinant: the NS1 protein four C-terminal residues modulate pathogenicity. *Proc Natl Acad Sci U S A* 2008, 105(11):4381-4386.

13 Katz JM, Lu X, Tumpey TM, Smith CB, Shaw MW, Subbarao K: Molecular correlates of influenza A H5N1 virus pathogenesis in mice. *J Virol* 2000, 74(22):10807-10810. 10.1128/JVI.74.22.10807-10810.2000

14 Kawaoka Y, Webster RG: Sequence requirements for cleavage activation of influenza virus hemagglutinin expressed in mammalian cells.*Proc Natl Acad Sci U S A* 1988, 85(2):324-328. 10.1073/pnas.85.2.324

15 Kimura M: A simple method for estimating evolutionary rates of base substitutions through comparative studies of nucleotide sequences. *J Mol Evol* 1980, 16(2):111-120. 10.1007/BF01731581

16 Kobasa D, Wells K, Kawaoka Y: Amino acids responsible for the absolute sialidase activity of the influenza A virus neuraminidase: relationship to growth in the duck intestine. *J Virol* 2001, 75(23):11773-11780.

17 Manzoor R, Sakoda Y, Nomura N, Tsuda Y, Ozaki H, Okamatsu M, Kida H: PB2 protein of a highly pathogenic avian influenza virus strain A/chicken/Yamaguchi/7/2004 (H5N1) determines its replication potential in pigs. *J Virol* 2009, 83(4):1572-1578.

18 Matrosovich M, Zhou N, Kawaoka Y, Webster R: The surface glycoproteins of H5 influenza viruses isolated from humans, chickens, and wild aquatic birds have distinguishable properties. *J Virol* 1999, 73(2):1146-1155.

19 Matrosovich MN, Krauss S, Webster RG: H9N2 influenza A viruses from poultry in Asia have human virus-like receptor specificity. *Virology*2001, 281(2):156-162.

20 Mishra AC, Cherian SS, Chakrabarti AK, Pawar SD, Jadhav SM, Pal B, Raut S, Koratkar S, Kode SS: A unique influenza A (H5N1) virus causing a focal poultry outbreak in 2007 in Manipur, India. *Virol J* 2009, 6: 26.

21 Monne I, Fusaro A, Al-Blowi MH, Ismail MM, Khan OA, Dauphin G, Tripodi A, Salviato A, Marangon S, Capua I, Cattoli G: Co-circulation of two sublineages of HPAI H5N1 virus in the Kingdom of Saudi Arabia with unique molecular signatures suggesting separate introductions into the commercial poultry and falconry sectors. *J Gen Virol* 2008, 89(Pt 11):2691-2697.

22 Naffakh N, Tomoiu A, Rameix-Welti MA, van der Werf S: Host restriction of avian influenza viruses at the level of the ribonucleoproteins.*Annu Rev Microbiol* 2008, 62: 403-424.

23 Nagarajan S, Rajukumar K, Tosh C, Ramaswamy V, Purohit K, Saxena G, Behera P, Pattnaik B, Pradhan HK, Dubey SC: Isolation and pathotyping of H9N2 avian influenza viruses in Indian poultry. *Vet Microbiol* 2009, 133(1-2):154-163.

24 Obenauer JC, Denson J, Mehta PK, Su X, Mukatira S, Finkelstein DB, Xu X, Wang J, Ma J, Fan Y, Rakestraw KM, Webster RG, Hoffmann E, Krauss S, Zheng J, Zhang Z, Naeve CW: Large-scale sequence analysis of avian influenza isolates. *Science* 2006, 311(5767):1576-1580.

25 Ohuchi M, Ohuchi R, Feldmann A, Klenk HD: Regulation of receptor binding affinity of influenza virus hemagglutinin by its carbohydrate moiety. *J Virol* 1997, 71(11):8377-8384.

26 Pawar S, Chakrabarti A, Cherian S, Pande S, Nanaware M, Raut S, Pal B, Jadhav S, Kode S, Koratkar S, Thite V, Mishra A: An avian influenza A(H11N1) virus from a wild aquatic bird revealing a unique Eurasian-American genetic reassortment. *Virus Genes* 2010, 41(1):14-22.

27 Pawar SD, Kale SD, Rawankar AS, Koratkar SS, Raut CG, Pande SA, Mullick J, Mishra AC: Avian influenza surveillance reveals presence of low pathogenic avian influenza viruses in poultry during 2009-2011 in the West Bengal State, India. *Virol J* 2012, 9: 151.

28 Rao BL: Influenza surveillance in Pune (India) during 1983-85. *Indian J Med Res* 1987, 85: 492-495.

29 Schmieder R, Edwards R: Quality control and preprocessing of metagenomic datasets. *Bioinformatics* 2011, 27(6):863-864.

30 Schnell JR, Chou JJ: Structure and mechanism of the M2 proton channel of influenza A virus. *Nature* 2008, 451(7178):591-595.

31 Shaw M, Cooper L, Xu X, Thompson W, Krauss S, Guan Y, Zhou N, Klimov A, Cox N, Webster R, Lim W, Shortridge K, Subbarao K: Molecular changes associated with the transmission of avian influenza a H5N1 and H9N2 viruses to humans. *J Med Virol* 2002, 66(1):107-114. 10.1002/jmv.2118

32 Tamura K, Peterson D, Peterson N, Stecher G, Nei M, Kumar S: MEGA5: molecular evolutionary genetics analysis using maximum likelihood, evolutionary distance, and maximum parsimony methods. *Mol Biol Evol* 2011, 28(10):2731-2739.

33 Thomas M, Kranjec C, Nagasaka K, Matlashewski G, Banks L: Analysis of the PDZ binding specificities of Influenza A virus NS1 proteins. *Virol J* 2011, 8: 25.

34 Tosh C, Nagarajan S, Behera P, Rajukumar K, Purohit K, Kamal RP, Murugkar HV, Gounalan S, Pattnaik B, Vanamayya PR, Pradhan HK, Dubey SC: Genetic analysis of H9N2 avian influenza viruses isolated from India. *Arch Virol* 2008, 153(8):1433-1439.

35 Virmani N, Bera BC, Singh BK, Shanmugasundaram K, Gulati BR, Barua S, Vaid RK, Gupta AK, Singh RK: Equine influenza outbreak in India (2008-09): virus isolation, sero-epidemiology and phylogenetic analysis of HA gene. *Vet Microbiol* 2010, 143(2-4):224-237.

36 Weis W, Brown JH, Cusack S, Paulson JC, Skehel JJ, Wiley DC: Structure of the influenza virus haemagglutinin complexed with its receptor, sialic acid. *Nature* 1988, 333(6172):426-431.

# CHAPTER 8

# Risk Factors for Infectious Diseases in Backyard Poultry Farms in the Poyang Lake Area, China

**Yong Wang,[1] Zhiben Jiang,[2] Zhenyu Jin,[2] Hua Tan,[2] and Bing Xu[1,2,3,*]**

1 School of Environment, Tsinghua University, Beijing, China

2 College of Global Change and Earth System Science, Beijing Normal University, Beijing, China

3 Department of Geography, University of Utah, Salt Lake City, Utah, United States of America

## ABSTRACT

Emergence and transmission of infectious diseases have an enormous impact on the poultry industry and present a serious threat to the health of humans and wild birds. Noncommercial poultry operations, such as backyard poultry facilities in China, are potential sources of virus exchange between commercial poultry and wild birds. It is particularly critical in wetland areas where backyard poultry have close contact with commercial poultry and migratory birds, therefore increasing the risk of contracting infectious diseases. To evaluate the transmission risks, a cross-sectional study was undertaken in the Poyang Lake area, China, involving 309 residents in the backyard poultry farms in three counties

(Region A, B, and C) of Jiangxi Province. We examined the backyard poultry population, poultry species, presence of poultry deaths from infectious diseases, food sources, and biosecurity practices. Region B ranked highest for biosecurity while region C ranked lowest. The risks of infectious diseases were assessed by adjusted odds ratio based on multivariate logistic regression analysis. Potential risk factors in the three regions of the study site were compared. In Region A, significant factor was contact of poultry with wild birds (OR: 6.573, 95% CI: 2.148–20.115, P=0.001). In Region B, the most significant factor was contact of poultry with neighboring backyard waterfowls (OR: 3.967, 95% CI: 1.555–10.122, P=0.004). In Region C, significant factors were poultry purchase from local live bird markets (OR: 3.740, 95% CI: 1.243–11.255, P=0.019), and contact of poultry with wild birds (OR: 3.379, 95% CI: 1.058–10.791, P=0.040). In summary, backyard poultry was significantly affected by neighboring commercial poultry and close contact with wild birds. The results are expected to improve our understanding of the transmission risks of infectious diseases in a typical backyard poultry environment in rural China, and address the need to improve local farming practices and take preventive measures.

# 1. INTRODUCTION

China is one of the world primary producers of poultry products, and poultry is one of the main sources of meat in the country [1]. Poultry production generates 16 million tons of meat and 27 million tons of eggs annually in China [1]. The economic impact of infectious diseases on the poultry industry is significant. In 2004, the total loss caused by the highly pathogenic avian influenza virus (HPAIV) H5N1 was 4.5 billion US dollars in China [2]. Husbandry practices including biosecurity measures, poultry species and rearing scale vary between different poultry raising systems [3], and understanding these practices is important for implementing preventive measures.

Therefore, risk factors that affect the dynamics of infectious disease transmission between different birds and across bird species, need to be identified. Wild waterfowl is recognized as a natural host for influenza A viruses [4–6]. The interaction of domestic poultry with other animals, particularly wild birds, has been recognized as a possible source of avian disease in domestic flocks [7]. It may also allow the disease to be transmitted from domestic poultry back to the wild bird populations. Some production practices, such as free-ranging may increase the chance of such an interaction and therefore the risk of disease transmission [8,9]. Although the role that migratory birds play in the transmission of the current HPAIV H5N1 to domestic poultry is unclear, phylogenetic comparison of the viruses obtained from different hosts suggest that migratory birds carry the virus and spread it to domestic birds [10,11].

The transportation of poultry or working utensils may facilitate the transmission and spread of avian diseases [12,13]. In particular, transportation via live bird markets is believed to have played an instrumental role in several outbreaks of avian diseases that have caused significant economic loss in the United States [14–16]. The spread of avian diseases has also been associated with the movement of humans between different bird flocks [17–19].

Husbandry practices in China's poultry-producing systems have not been fully examined in previous studies. From our observation, there are two distinct systems for commercial production and backyard poultry keeping. Local veterinarian told us that commercial farms usually operate on a large scale and poultry is generally kept in high density and enclosed housing conditions with high biosecurity measures. On the contrary, the majority of backyard poultry owners raises the poultry in open yards in a small scale, and do not apply much biosecurity measures. The flocks are raised for owners' consumption only.

The HPAIVs that have been isolated from backyard flocks [16,20] and free range poultry are usually associated with the transmission of the lowly pathogenic avian influenza virus from wild birds [7,21]. The exotic Newcastle disease outbreak that occurred between 2002 and 2003 in the United States started among backyard poultry flocks and subsequently spread to commercial flocks [22]. An improved understanding of the risk factors for the backyard operation will help to control the transmission of infectious diseases and implement preventive strategies.

Poyang Lake area was selected as the study area because it is the largest freshwater lake in China with a total water area of more than 4,000 km$^2$ in the summer wet season and less than 300 km$^2$ during winter dry months [23]. Over 520,000 wild birds were found at the Poyang Lake during the wintering period in 2006 [24].

The objectives of this epidemiological survey were to acquire information on the biosecurity procedures adopted by backyard poultry growers, the number and type of birds raised, and the husbandry practices; and to identify risk factors that are associated with infections in backyard poultry in rural China.

# 2. MATERIALS AND METHODS

## 2.1 Ethics statement

The protocol of the study was approved by the institutional ethics committee in the School of Environment, Tsinghua University. Prior to interview, a printed document detailing objective and format of the study was given to each participant. Signed consent was obtained from all participants prior to interview.

## 2.2 Survey locations

Three counties were selected around Poyang Lake for the study (Figure 1): Nanchang County (Region A), De'an County (Region B) and Duchang County (Region C). These three regions were chosen for a comparative study because they represented different backyard poultry farming with different economic development and presence of commercial poultry industry.

Region A is located in the southern part of Poyang Lake, and close to the Nanji Wetlands National Nature Reserve. The annual net income of local farmers was 1,182 USD per person in 2010. Until the end of 2010, the whole county had 8.5 million laying ducks, ranking the highest in the Poyang Lake area [25]. Region B is located in the western part of Poyang Lake, and close to the Poyang Lake National Nature Reserve. The annual net income was 909 USD per person in 2010. Until the end of 2010, the whole county had 4 million laying hens, ranking the highest in the

Poyang Lake area [25]. Region C is located in the northern part of Poyang Lake, and close to the Duchang Migratory Birds Provincial Nature Reserve. The annual net income of farmers was 554 USD per person in 2010 [25]. There was no commercial farm with more than 300 birds in this area according to the local veterinary services,

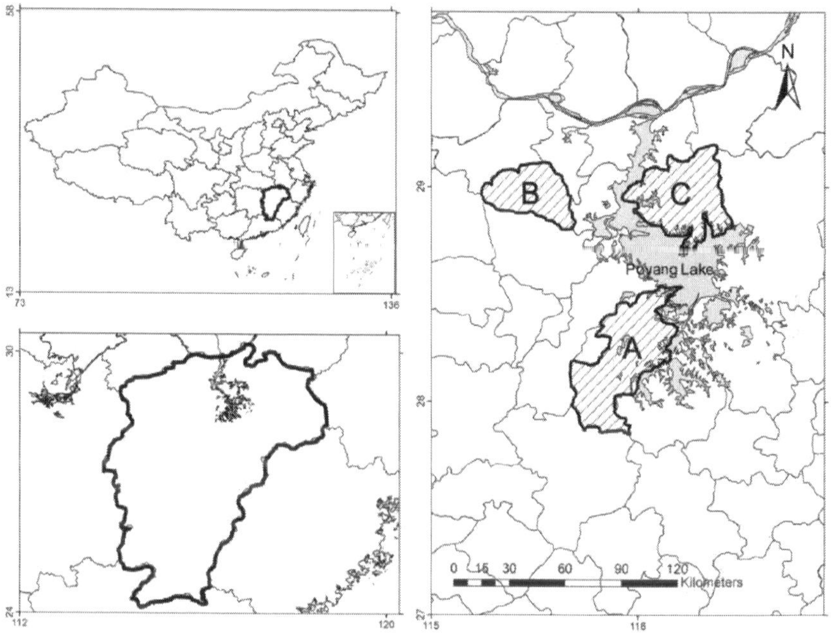


**Figure 1** Study area around Poyang Lake of Jiangxi Province.

## 2.3 Questionnaires

A cross-sectional survey was conducted to investigate rearing and biosecurity practices, and flock health history. A draft questionnaire was designed and pretested at thirteen farms. This draft was modified based on the results obtained from the pretest. The translated questionnaire is provided in the supporting information (Text S1).

The final questionnaire consisted of three sections. Section 1 referred to the demographic features and characteristics of the poultry species and farm size. Section 2 involved biosecurity practices of the household including poultry raising style, routine trading behavior, vaccination procedure, and contact with wild birds. Section 3 acquired information on flock health history. In the questionnaire, "contact with neighboring poultry" is defined as sharing foraging with neighboring flocks. "Accessibility to wild birds" is defined as presence of wild birds in the activity space of poultry except for sharing feed. "Contact with wild birds" is defined as sharing foraging with wild birds. "Free ranging" indicates no restriction on flocks' foraging activity in the daytime.

Before the survey was conducted among the residents in the villages, we exchanged information with local veterinarians in the county government and familiarized ourselves with the basic conditions of backyard poultry rearing. The local government would organize free immunizations twice a year. The only vaccine that local residents are required to give the backyard flocks is H5 subtype avian influenza. Local veterinarians pointed out that the prevailing infectious diseases were avian influenza, Newcastle disease, *Escherichia coli*infection, fowl cholera and egg drop syndrome. In addition, several strains of avian influenza virus and Newcastle disease virus have been previously isolated from the backyard poultry flocks in the local area. When multiple unexplained deaths of backyard poultry were reported to the local government, veterinarians would be sent to conduct a necropsy to determine cause of death and advise appropriate control measures.

## 2.4 Data collection

Region A was first divided into 18 townships, each consisting of 10–31 administrative villages, and each administrative village comprised several natural villages. The addresses of the commercial and backyard poultry farms were collected from local livestock offices. Based on this information, we then chose three townships with the highest density of commercial poultry farming. Within each administrative village, two natural villages were randomly sampled. Every natural village was assigned with a unique number. In each selected natural village, the number of households was finally chosen in proportion to the number of poultry-raising farms. A total of 83 backyard farms were included in Region A. Similarly, 157 backyard farms were selected for Region B. There was no commercial poultry farming in Region C, so the two townships in Region C were chosen randomly. A total of 69 backyard farms were selected from Region C by the same sampling method used in Region A. Backyard poultry owners were initially contacted by local village leader, informed of the purpose and format of the study, and made appointments for interviews. If the poultry owners were absent, the questionnaires would be completed by an adult ($\geq$ 18 years) family member who helped with raising the poultry. The questionnaire was administered by three trained individuals who interviewed farm owners or an adult family member if the owner was absent. Among a total of 309 questionnaires, only 15 were not completed by poultry owners. The questionnaires were maintained by the investigators and would not be released to the third party for confidentiality.

## 2.5 Quality control and statistical analysis

The questionnaire data were checked, errors in data entry were corrected, and completeness was evaluated. To cross check the data entry, two individual interviewers entered twice the data into EpiData V3.1 (EpiData, Odense, Denmark) and then analyzed them in S+ 8.2 (TIBCO, Palo Alto, CA, USA). Data distributions were examined across study strata, including density of commercial poultry farms and species of commercial poultry. Region A differs from Region B in that the major commercial poultry in Region A is free ranging laying ducks, while the major commercial poultry in Region B is captive laying hens. Therefore, Region A is compared with Region B in terms of species of commercial poultry and rearing style. Region A+B differs from Region C in that both Region A and Region B own large scale commercial poultry farms in high densities, while Region C does not own any large scale commercial poultry farms. Therefore, Region A+B is compared with Region C in terms of scale or density of commercial poultry farms. Mann–Whitney tests were performed to compare between regions for ranking data. chi-square ($\chi^2$) or Fisher's exact tests were performed for counting data obtained from survey.

Flock health history data were acquired to design a case–control study. Univariate analysis was conducted to test for associations between presence of disease mortality and health history by using $\chi^2$ tests for categorical variables. To examine the independence of explanatory variables, multivariate logistic regression analysis was performed. Variables from univariate analyses with $P$ values < 0.10 were included in the multivariate regression. The final model of potential risk factors was constructed by stepwise backward selection applying the iterative maximum-likelihood estimation procedure. Adjusted odds ratios (ORs) and 95% confidence intervals (CIs) were calculated. A $P$ value < 0.05 was considered statistically significant.

## 3. RESULTS

A total of 62 villages in Poyang Lake area were included in this study. Three hundred and nine backyard poultry owners completed their questionnaires.

### 3.1 Characteristics of backyard poultry flocks

A majority of families owned chicken only (82.20%) in their backyards. Some households raised more than one species of poultry, for example, chicken and duck (13.27%). Most flock size was less than 10 for chicken (161/233), duck (38/49) and goose (6/7). Most of the growers raised backyard birds as a source of food for family use, and only 2.91% of growers (6.02% of growers in Region A, 2.55% of growers in Region B, and none in Region C) reported selling their birds to dealers in all three regions.

The data summarized in Table 1 illustrate responses to questions pertaining to the flock characteristics of three regions, including species, number and proportion of poultry. There was a significant difference between Region A and Region B in terms of species of poultry raised in the backyard ($P < 0.001$). The percentage of chickens was only 51.81% in Region A versus 91.10% in Region B. The percentage of farmers who raised both chickens and ducks was 38.55% in Region A, while it was only 5.70% in Region B. Comparison between Regions A+B and Region C is as follows: 77.50% of the farms in Regions A+B raised chickens alone, while this percentage was 98.55% in Region C; and 17.08% of the farms in Regions A+B raised duck and chickens but no farms did this in

Region C. There were significant differences in chicken capacity per farm between Region A and Region B ($P = 0.014$). The capacity was < 10 chicken in 44.16% of farms in Region A versus 59.62% in Region B. Only two chicken farms had > 50 birds and two duck farms had > 25 birds in three regions. Raising geese was not common (three farms in Region A, four in Region B, and one in Region C).

**Table 1** Descriptive analysis of species, number, purpose and proportion (%) of poultry included in survey backyard farms in Jiangxi, China.

| | A+B vs C[a] | | | A vs B[b] | | |
|---|---|---|---|---|---|---|
| Characteristic | A+B (%) | C (%) | P[c] | A (%) | B (%) | P[c] |
| **Species of poultry*** | n=240 | n=69 | <0.001 | n=83 | n=157 | <0.001 |
| Chicken only | 186 (77.50) | 68 (98.55) | | 43 (51.81) | 143 (91.10) | |
| Duck only | 6 (2.50) | 0 (0.00) | | 5 (6.00) | 1 (0.60) | |
| Goose only | 0 (0.00) | 0 (0.00) | | 0 (0.00) | 0 (0.00) | |
| Chicken +duck | 41 (17.08) | 0 (0.00) | | 32 (38.55) | 9 (5.70) | |
| Chicken +goose | 5 (2.08) | 1 (1.45) | | 2 (2.41) | 3 (1.90) | |
| Duck +goose | 1 (0.42) | 0 (0.00) | | 1 (1.20) | 0 (0.00) | |
| Chicken +duck +goose | 1 (0.42) | 0 (0.00) | | 0 (0.00) | 1 (0.60) | |
| **Number of birds*** | | | | | | |
| Chicken | n=233 | n=69 | 0.830 | n=77 | n=156 | 0.014 |
| 1-10 | 127 (54.51) | 34 (49.28) | | 34 (44.16) | 93 (59.62) | |
| 11-25 | 83 (35.62) | 33 (47.83) | | 31 (40.26) | 52 (33.33) | |
| 26-50 | 21 (9.01) | 2 (2.90) | | 12 (15.58) | 9 (5.77) | |
| 51-100 | 2 (0.86) | 0 (0.00) | | 0 (0.00) | 2 (1.28) | |

| | | | | | | |
|---|---|---|---|---|---|---|
| Duck | n=49 | n=0 | | n=38 | n=11 | 0.046 |
| 1-10 | 38 (77.55) | 0 (0.00) | | 27 (71.05) | 11 (100.00) | |
| 11-25 | 9 (18.37) | 0 (0.00) | | 9 (23.68) | 0 (0.00) | |
| 26-50 | 2 (4.08) | 0 (0.00) | | 2 (5.26) | 0 (0.00) | |
| Goose | n=7 | n=1 | 0.705 | n=3 | n=4 | 0.248 |
| 1-10 | 6 (85.71) | 1 (100.00) | | 2 (66.67) | 4 (100.00) | |
| 11-25 | 1 (14.29) | 0 (0.00) | | 1 (33.33) | 0 (0.00) | |
| **Purpose of raising**** | n=240 | n=69 | | n=83 | n=157 | |
| Source of food | 240 (100.00) | 69 (100.00) | 1.000 | 83 (100.00) | 157 (100.00) | 1.000 |
| Source of income | 9 (3.75) | 0 (0.00) | 0.103 | 5 (6.02) | 4 (2.55) | 0.178 |

[a] Comparing penned commercial poultry (Regions A+B) versus Region C with no commercial poultry. [b] Comparing penned laying ducks (Region A) versus penned laying hens (Region B). [c] $\chi^2$ or Fisher's exact tests. * Exclusive. ** Non-exclusive. Individual question is annotated with exclusive or non-exclusive. If the question is exclusive, the percentages for the answers will sum up to 100%. If the question is non-exclusive, the percentages for the answers will not sum up to 100%.

## 3.2 Contact between backyard poultry and other birds

Laying duck owners will always buy younger poultry and poultry feed from live bird market. This may promote indirect interaction between their backyard poultry and the live bird market. However, a large number of farmers (90.61%) did not buy any food from wet market for their backyard poultry. Wet market and hatching themselves were two main sources of younger poultry, and only 18 (5.83%) owners reported that their younger poultry were obtained from other sources, such as commercial laying farms. There were 220 (71.20%) owners who reported presence of wild birds in the activity space of their backyard poultry. and 139 (44.98%) owners who reported that their backyard poultry shared foraging with wild birds. In addition, 242 (78.32%) owners reported that their backyard poultry shared foraging with neighboring poultry flocks.

Table 2 summarizes the answers to questions regarding the direct contact, accessibility of backyard poultry with other birds, and indirect disease transmission sources, such as live bird market. There was no obvious difference in the source of younger poultry between the regions, except that the growers obtained younger domestic birds from the market more frequently in Region A than in Region B; similarly, it was more frequent in Regions A+B than in Region C ($P$=0.006). Likewise, the growers obtained feed for poultry from the market more frequently in Region A than in Region B ($P$=0.013). Similarly, in Regions A+B growers obtained poultry feed from the market more frequently than in Region C ($P$=0.002). The contact of backyard flocks in Region C with their neighboring chickens, 91.30%, was more frequent than that in Regions A+B (74.58%, $P$ = 0.003), while the contact of backyard flocks in Region C with their neighboring waterfowls, 2.9%, was less frequent than that in Regions A+B (18.75%, $P$ = 0.001). Wild birds were frequently observed in all three regions: 67.47% in Region A; 71.97% in Region B; and 73.91% in Region C, and there was no significant difference between these regions. In Region A, households reported a higher frequency of poultry contact with wild birds (73.21%) than that in Region B (49.59%, $P$ = 0.004). And there was also a significant difference between Region C (82.35%) and Regions A+B (57.40%, $P$ = 0.001).

## 3.3 Disease prevention measures and farming style

Among backyard flocks, 44.66% were not immunized. Among 171 vaccinated flocks, 68.42% were immunized by the government. There were 164 flocks that all poultry in their backyards completed their immunization, while only seven flocks were left for partial vaccination (9.3% in Region A, 2.91% in Region B, and none in Region C). Growers explained that vaccination might affect egg production. A great number of owners (88.35%) raise backyard poultry every year, and 89.97% of the owners kept their poultry in a free ranging style.

There was no difference in the frequency of vaccination between Region A and Region B, but the growers in Regions A+B reported more frequent vaccination than those in Region C (Table 3). The vaccination rate in

Regions A+B was significantly higher than in Region C ($P<0.001$). In Region B, growers reported a higher frequency of vaccination by government (87.38%), while in Region A, vaccination by government was uncommon, and a significantly larger number of growers reported that they vaccinated their birds by themselves (86.05%, $P<0.001$). The households raised backyard poultry more frequently every year in Region A (96.39%) than those in Region B did (81.89%, $P=0.001$), and there was also a difference between Regions A+B (86.25%) and Region C (95.65%, $P=0.032$). There was a significant difference in the proportion of free ranging between Region A (96.39%) and Region B (88.54%, $P=0.041$), but no significant difference between Regions A+B and Region C ($P=0.162$).

**Table 2** The source of younger poultry and feed, and contact between backyard poultry and other birds.

| | A+B vs C | | | A vs B | | |
|---|---|---|---|---|---|---|
| Characteristic | A+B (%) | C (%) | *P* | A (%) | B (%) | *P* |
| **Younger poultry source**** | n=240 | n=69 | | n=83 | n=157 | |
| Wet market | 129 (53.75) | 24 (34.78) | 0.006 | 55 (66.27) | 74 (47.13) | 0.005 |
| Hatching themselves | 132 (55.00) | 41 (59.42) | 0.515 | 41 (49.40) | 91 (57.96) | 0.206 |
| **Feed source**** | n=240 | n=69 | | n=83 | n=157 | |
| Wet market | 29 (12.08) | 0 (0.00) | 0.002 | 16 (19.28) | 13 (8.28) | 0.013 |
| Made by themselves | 238 (99.17) | 69 (100.00) | 0.448 | 81 (97.59) | 157 (100.00) | 0.051 |
| **Contact with neighboring poultry**** | n=240 | n=69 | | n=83 | n=157 | |
| Chicken | 179 (74.58) | 63 (91.30) | 0.003 | 57 (68.67) | 122 (77.71) | 0.127 |
| Waterfowl | 45 (18.75) | 2 (2.90) | 0.001 | 20 (24.10) | 25 (15.92) | 0.124 |
| None | 61 (25.42) | 6 (8.70) | 0.003 | 26 (31.33) | 35 (22.29) | 0.127 |
| **Accessibility to wild birds*** | n=240 | n=69 | 0.573 | n=83 | n=157 | 0.468 |
| Yes | 169 (70.42) | 51 (73.91) | | 56 (67.47) | 113 (71.97) | |

| | | | | | | |
|---|---|---|---|---|---|---|
| No | 71 (29.58) | 18 (26.09) | | 27 (32.53) | 44 (28.03) | |
| **Contact with wild birds*** | n=169 | n=51 | 0.001 | n=56 | n=113 | 0.004 |
| Yes | 97 (57.40) | 42 (82.35) | | 41 (73.21) | 56 (49.56) | |
| No | 72 (42.60) | 9 (17.65) | | 15 (26.79) | 57 (50.44) | |

* Exclusive. ** Non-exclusive. Individual question is annotated with exclusive or non-exclusive. If the question is exclusive, the percentages for the answers will sum up to 100%. If the question is non-exclusive, the percentages for the answers will not sum up to 100%.

**Table 3** Characteristics of vaccination and raising practice.

| | A+B vs C | | | A vs B | | |
|---|---|---|---|---|---|---|
| **Characteristic** | **A+B (%)** | **C (%)** | *P* | **A (%)** | **B (%)** | *P* |
| **Frequency of vaccination*** | n=240 | n=69 | <0.001 | n=83 | n=157 | 0.172 |
| 0 | 94 (39.17) | 44 (63.77) | | 40 (48.19) | 54 (34.39) | |
| 1 | 60 (25.00) | 18 (26.09) | | 20 (24.10) | 40 (25.48) | |
| 2 | 69 (28.75) | 7 (10.14) | | 9 (10.84) | 60 (38.22) | |
| ≥3 | 17 (7.08) | 0 (0.00) | | 14 (16.87) | 3 (1.91) | |
| **Vaccinated by*** | n=146 | n=25 | 0.071 | n=43 | n=103 | <0.001 |
| Government | 96 (65.75) | 21 (84.00) | | 6 (13.95) | 90 (87.38) | |
| Poultry owner | 50 (34.25) | 4 (16.00) | | 37 (86.05) | 13 (12.62) | |
| **Poultry vaccinated*** | n=146 | n=25 | 0.265 | n=43 | n=103 | 0.101 |
| All | 139 (95.21) | 25 (100.00) | | 39 (90.70) | 100 (97.09) | |
| Partly | 7 (4.79) | 0 (0.00) | | 4 (9.30) | 3 (2.91) | |
| **Raising frequency*** | n=240 | n=69 | 0.032 | n=83 | n=157 | 0.001 |

| | | | | | | |
|---|---|---|---|---|---|---|
| Every year | 207 (86.25) | 66 (95.65) | | 80 (96.39) | 127 (80.89) | |
| Some years | 33 (13.75) | 3 (4.35) | | 3 (3.61) | 30 (19.11) | |
| **Type of raising*** | n=240 | n=69 | 0.162 | n=83 | n=157 | 0.041 |
| In captivity | 21 (8.75) | 10 (14.49) | | 3 (3.61) | 18 (11.46) | |
| Free range | 219 (91.25) | 59 (85.51) | | 80 (96.39) | 139 (88.54) | |

* Exclusive. Individual question is annotated with exclusive or non-exclusive. If the question is exclusive, the percentages for the answers will sum up to 100%. If the question is non-exclusive, the percentages for the answers will not sum up to 100%.

## 3.4 Health history

Among the surveyed participants, 48.87% of owners reported backyard poultry mortality caused by infectious diseases in the past years. Eighty four owners reported deaths in the past 12 months (during the year of 2009), and all of the 84 owners also reported deaths in the previous 12 months (during the year before 2009). The growers in Region A reported a higher frequency of infectious disease (59.04%) than those in Region B (39.49%, $P$=0.004), but there was no significant difference between Regions A+B and Region C (Table 4).

## 3.5 Potential risk factors

Characteristics of backyard poultry rearing practices differ in three regions. Region A owns the most free-ranging laying ducks, with poor completion of required vaccination. The most common source for younger poultry is from the live bird market. The backyard poultry in this region has relatively high contact rate with wild birds and neighboring waterfowls. Region B has the largest scale of commercial production of laying hens with highest biosecurity measures, such as regular disinfection, timely vaccination, monitored transportation of products, etc. Region C has the lowest annual income per person with worst vaccination completion. Backyard poultry in this region has the highest contact rate with neighboring poultry and wild birds, and highest accessibility to wild birds. The poultry mortality caused by infectious diseases ranks the lowest in region B and high in both Region A and C. Implications for different disease control strategies due to regional differences will be suggested and discussed in terms of these known factors for disease spread. These known factors are considered as potential risk factors for infectious diseases in backyard poultry in three different regions and analyzed through univariate analysis by a case control study design.

**Table 4** Health history in the past two years.

| | A+B vs C | | | A vs B | | |
|---|---|---|---|---|---|---|
| Characteristic | A+B (%) | C (%) | *P* | A (%) | B (%) | *P* |
| Infectious disease death* | n=240 | n=69 | 0.087 | n=83 | n=157 | 0.004 |
| Yes | 111 (46.25) | 40 (57.97) | | 49 (59.04) | 62 (39.49) | |
| No | 129 (53.75) | 29 (42.03) | | 34 (40.96) | 95 (60.51) | |
| The onset of infection disease death* | n=111 | n=40 | 0.308 | n=49 | n=62 | 0.986 |
| Death occurs only in the last 12 months | 0 (0.00) | 0 (0.00) | | 0 (0.00) | 0 (0.00) | |
| Death occurs only in 12 months ago | 52 (46.85) | 15 (37.50) | | 23 (46.94) | 29 (46.77) | |
| Death occurs in both time | 59 (53.15) | 25 (62.50) | | 26 (53.06) | 33 (53.23) | |

*Exclusive. Individual question is annotated with exclusive or non-exclusive. If the question is exclusive, the percentages for the answers will sum up to 100%. If the question is non-exclusive, the percentages for the answers will not sum up to 100%.

Logistic regression was applied to model backyard poultry deaths caused from infectious diseases in 2009 and potential risk factors. Table 5 shows the results of the univariate analyses in Region A. In this region, "poultry contact with wild birds" (OR: 5.714, 95% CI: 1.981–16.481, $P$=0.001); and "poultry contact with neighboring backyard waterfowls" (OR: 2.937, 95% CI: 1.034–8.345, $P$=0.043) were significantly associated with infection ($P \leq 0.05$) by univariate analysis. Table 6 shows the results of the univariate analyses in Region B. In this region, the factors positively associated with case farms by univariate analysis were: "raising more than one kind of poultry" (OR: 5.295, 95% CI: 1.643–17.063, $P$=0.005); "poultry contact with neighboring backyard waterfowls" (OR: 4.879, 95% CI: 1.959–12.155, $P$=0.001); and "farm accessible to wild birds" (OR: 3.452, 95% CI: 1.136–10.488, $P$=0.029). Three variables were considered for logistic regression. Table 7 shows the results of the univariate analyses in Region C. In this region, "buy younger poultry from wet market" (OR: 3.250, 95% CI: 1.149–9.193, $P$=0.026); and "poultry contact with wild birds" (OR: 2.891, 95% CI: 0.970–8.616, $P$=0.057) were the two significant risk factors by univariate analysis.

Multivariate analysis of potential risk factors was performed combining all three regions (Table 8). In Region A, six variables with $P \leq 0.1$ were considered for inclusion in the final multivariate logistic regression model to estimate independence of effects. The final logistic regression model included one variable as an independent risk factor for disease death: "poultry contact with wild birds" (OR: 6.573, 95% CI: 2.148–20.115, $P$=0.001). In Region B, "poultry contact with neighboring backyard waterfowls" was independently and positively associated with infection (OR: 3.967, 95% CI: 1.555–10.122, $P$=0.004). In Region C, two factors were identified as independent risk factors in the final model: "buy younger poultry from wet market" (OR: 3.740, 95% CI: 1.243–11.255, $P$=0.019); and "poultry contact with wild birds" (OR: 3.379, 95% CI: 1.058–10.791, $P$=0.040).

**Table 5** Univariate analysis of potential risk factors for infectious disease in backyard poultry in Region A (Nanchang Country), Poyang Lake area, Jiangxi, China, 2009.

| Characteristic | Case farms n=26, n (%) | Control farms n=57, n (%) | OR (95% CI)* | P value |
|---|---|---|---|---|
| Raising more than one species of poultry | 12 (46.15) | 23 (40.35) | 1.267 (0.497-3.228) | 0.602 |
| Free range | 25 (96.15) | 55 (96.49) | 0.909 (0.079-10.449) | 0.939 |
| Buy younger poultry from wet market | 10 (73.08) | 36 (63.16) | 1.583 (0.571-4.391) | 0.377 |
| Buy poultry food from wet market | 2 (7.69) | 14 (24.56) | 0.256 (0.054-1.222) | 0.088 |
| Raising for income | 2 (7.69) | 3 (5.26) | 1.500 (0.235-9.565) | 0.668 |
| Vaccinated by government | 4 (15.38) | 2 (3.51) | 5.000 (0.853-29.293) | 0.074 |
| Vaccinated by poultry owners | 7 (26.92) | 30 (52.63) | 0.332 (0.121-0.911) | 0.032 |
| All vaccinated | 10 (38.46) | 29 (50.88) | 0.603 (0.234-1.553) | 0.295 |
| Partly vaccinated | 1 (3.85) | 3 (5.26) | 0.720 (0.071-7.270) | 0.781 |
| Poultry contact with neighboring backyard chickens | 21 (80.77) | 36 (63.16) | 2.450 (0.804-7.463) | 0.115 |
| Poultry contact with neighboring backyard waterfowls | 10 (38.46) | 10 (17.54) | 2.937 (1.034-8.345) | 0.043 |
| Farm accessibility to wild birds | 21 (80.77) | 35 (61.40) | 2.640 (0.869-8.023) | 0.087 |
| Poultry contact with wild birds | 20 (76.92) | 21 (36.84) | 5.714 (1.981-16.481) | 0.001 |

* OR, odds ratio; CI, confidence interval

**Table 6** Univariate analysis of possible risk factors for infectious disease in backyard poultry in Region B (De'an Country), Poyang Lake area, Jiangxi, China, 2009.

| Characteristic | Case farms n=33, n (%) | Control farms n=124, n (%) | OR (95% CI) | *P* value |
|---|---|---|---|---|
| Raising more than one species of poultry | 7 (21.21) | 6 (4.84) | 5.295 (1.643-17.063) | 0.005 |
| Free range | 29 (87.88) | 110 (88.71) | 0.923 (0.282-3.015) | 0.894 |
| Buy younger poultry from wet market | 18 (54.55) | 53 (42.74) | 1.608 (0.743-3.479) | 0.228 |
| Buy poultry food from wet market | 3 (9.09) | 10 (8.06) | 1.140 (0.295-4.404) | 0.849 |
| Raising for income | 1 (3.03) | 3 (2.42) | 1.260 (0.127-12.527) | 0.843 |
| Vaccinated by government | 18 (54.55) | 72 (58.06) | 0.867 (0.400-1.876) | 0.717 |
| Vaccinated by poultry owners | 2 (6.06) | 11 (8.87) | 0.663 (0.140-3.148) | 0.605 |
| All vaccinated | 20 (60.61) | 80 (64.52) | 0.846 (0.384-1.863) | 0.678 |
| Partly vaccinated | 0 (0.00) | 3 (2.42) | Incalculable | NA |
| Poultry contact with neighboring backyard chickens | 30 (90.91) | 92 (74.19) | 3.478 (0.993-12.178) | 0.051 |
| Poultry contact with neighboring backyard waterfowls | 12 (36.36) | 13 (10.48) | 4.879 (1.959-12.155) | 0.001 |
| Farm accessibility to wild birds | 29 (87.88) | 84 (67.74) | 3.452 (1.136-10.488) | 0.029 |
| Poultry contact with wild birds | 13 (39.39) | 43 (34.68) | 1.224 (0.556-2.698) | 0.616 |

**Table 7** Univariate analysis of possible risk factors for infectious disease in backyard poultry in Region C (Duchang Country), Poyang Lake area, Jiangxi, China, 2009.

| Characteristic | Case farms n=25, n (%) | Control farms n=44, n (%) | OR (95% CI) | *P* value |
|---|---|---|---|---|
| Raising more than one species of poultry | 0 (0.00) | 1 (2.27) | Incalculable | NA |
| Free range | 20 (80.00) | 39 (88.64) | 0.513 (0.133-1.982) | 0.333 |
| Buy younger poultry from wet market | 13 (52.00) | 11 (25.00) | 3.250 (1.149-9.193) | 0.026 |
| Buy poultry food from wet market | 0 (0.00) | 0 (0.00) | Incalculable | NA |
| Raising for income | 0 (0.00) | 0 (0.00) | Incalculable | NA |
| Vaccinated by government | 7 (28.00) | 14 (31.82) | 0.833 (0.283-2.452) | 0.741 |
| Vaccinated by poultry owners | 0 (0.00) | 4 (9.09) | Incalculable | NA |
| All vaccinated | 7 (28.00) | 18 (40.91) | 0.562 (0.195-1.621) | 0.286 |
| Partly vaccinated | 0 (0.00) | 0 (0.00) | Incalculable | NA |
| Poultry contact with neighboring backyard chickens | 23 (92.00) | 40 (90.91) | 1.150 (0.195-6.773) | 0.877 |
| Poultry contact with neighboring backyard waterfowls | 1 (4.00) | 1 (2.27) | 1.792 (0.107-29.952) | 0.685 |
| Farm accessibility to wild birds | 19 (76.00) | 32 (72.73) | 1.187 (0.383-3.685) | 0.766 |
| Poultry contact with wild birds | 19 (76.00) | 23 (52.27) | 2.891 (0.970-8.616) | 0.057 |

**Table 8** Result of multivariate analysis of potential risk factor for backyard poultry death of infectious disease, Poyang Lake area, Jiangxi, China, 2009.

| Characteristic | Odds ratio | 95% CI | *P* value |
|---|---|---|---|
| **Region A** | | | |
| Poultry contact with wild birds | 6.573 | 2.148-20.115 | 0.001 |
| **Region B** | | | |
| Poultry contact with neighboring backyard waterfowls | 3.967 | 1.555-10.122 | 0.004 |
| **Region C** | | | |
| Buy younger poultry from wet market | 3.740 | 1.243-11.255 | 0.019 |
| Poultry contact with wild birds | 3.379 | 1.058-10.791 | 0.040 |

# 4. DISCUSSION AND CONCLUSION

A combination of descriptive and analytical epidemiological techniques was used to explore and evaluate potential risk factors associated with mortality of backyard poultry caused by infectious diseases in the Poyang Lake area of China.

It is summarized from the survey that the frequency of waterfowl raising in Region A is higher than that in Region B and Region C. Region A is in plain areas with dense river networks, which are conducive to the raising of waterfowls. Region B and Region C were in relatively hilly areas. Previous studies have recognized domestic ducks as a risk factor for disease occurrence owing to their potential role as a reservoir of infectious diseases [26–29]. The riparian zones of the lake provide major food sources including vegetation, fish and crustaceans for waterfowl. The water body can be contaminated by congregating sick ducks [30,31]. Backyard poultry may be exposed to virus transmission while they share the same water area.

Wet markets are considered to be an important source of infection [32]. Locally produced poultry is mixed with imported poultry for more than a day, thus viruses may be amplified in the markets [33]. Several subtypes of AIVs have previously been isolated from the markets in Poyang Lake area [34]. According to our interviews, no one purchased live adult poultry from markets and slaughtered them at home. A few people reported that they purchased live adult birds and had them slaughtered by the seller at the market several times annually. However, farmers do purchase younger poultry and poultry feed from live bird markets. Region A owns the most free-ranging laying ducks. The most common source for younger poultry is from the live bird market.

Viruses were found to be widely distributed among wild birds. Migratory waterfowl are particularly considered to be a natural reservoir of AIVs [5,35–37]. Resident birds have also been shown to harbor H5N1 AIVs [38,39]. Every winter, many migratory waterfowl inhabit along the riparian zones of Poyang Lake. This raises concern that migratory birds could introduce viruses into local resident wild birds, which might then deliver the virus to local flocks of domestic birds. Compared with those in Region B, the farmers in Region A and Region C reported higher frequency of backyard poultry contact with wild birds and neighboring poultry. The distance from Poyang Lake to Region B and the other two regions could be responsible for the difference in this risk factor.

Since 2005, H5N1 vaccination in China has been carried out by governmental veterinarians free of charge during two main seasons per year, March to April and October to November [40]. The difference in vaccination practices between Region A and other two regions showed that, self vaccination in Region A was organized to match with the production cycles of backyard poultry. However, the growers followed the governmental strategy of vaccinations twice in Region B and Region C. Our results indicate that vaccination and the timing of vaccination are important in making effective prevention strategies.

Most backyard poultry raised in the Poyang Lake area are in a free ranging style. This may increase the chance of poultry's being exposed to neighboring poultry and wild birds, which will in turn increase the disease risk of backyard poultry. Backyard poultry farming is scattered. It would be difficult to catch the actual chain infection by a single cross-sectional survey. This survey recorded occurrence of poultry deaths caused by infectious disease in the past years. Additionally, in our previous study, several virus strains (1 strain of H5 subtype, 4 strains of H9 subtype of avian influenza virus and 3 strains of Newcastle disease virus) among a total of 1340 samples collected from backyard poultry in this area, were isolated using specific pathogen free (SPF) chicken embryos (unpublished data).

To reduce potential bias, standardized questionnaires and trained interviewers were provided. It was recognized that recall bias was likely to occur, particularly for questions on the date of death of backyard poultry and date of vaccination each year. In our pilot survey, we found that special days, such as the Chinese New Year/Spring Festival, could help growers to recall events more easily. The interview time was chosen near the Spring Festival, and the growers were asked to recall events between two consecutive Spring Festivals (within the last 12 months) and out with the previous 12 months.

Three remaining factors including poultry contact with neighboring backyard waterfowl, poultry contact with wild birds, and buying younger poultry from wet markets, were important for the spread and maintenance of infectious diseases in Poyang Lake area. However, potential risk varies from region to region, as characteristics of backyard poultry rearing practices differ from region to region. Scale of commercial free ranging laying ducks in Region A is largest. Younger poultry of laying ducks are mainly purchased from live bird markets. It is easily spotted that wild birds share feeds with the free ranging laying ducks. There is no doubt that the backyard poultry will have close contact with wild birds in this region. It is hard to reduce the contacts except for raising less backyard poultry or reducing the scale of free ranging laying ducks. The poultry mortality caused by infectious diseases in Region B ranks the lowest, although it has the largest scale of commercial production of laying hens, which requires high biosecurity measures. Although the amount of backyard waterfowls is small, "poultry contact with neighboring backyard waterfowl" is a significant risk factor. Farmers may further reduce the backyard waterfowl breeding. The economic situation of farmers in Region C ranks the poorest with worst vaccination completion and no commercial farms. Farmers should reduce buying younger poultry from live bird market, and reduce the chance of poultry contact with wild birds. To enforce biosecurity measures and implementation of immunization by means of education may be a practical way of improving the overall poultry rearing practices, thus reducing disease risks of backyard poultry.

Overall, the owners should be advised to reduce free-range rearing of poultry. During feeding, the owner may prevent wild birds from joining the flock, and the remaining feed in the poultry activity area must be removed. Villagers should reduce buying younger poultry from wet markets. Controls and restrictions in the transportation of live poultry should be enforced, such as safe transport of live poultry through clean and disinfected transport vehicles and cages, at the currently available border area check points.

Our results showed that live bird markets and wild birds were two potential risk factors of infectious diseases for backyard poultry in the Poyang Lake area. This suggests that backyard poultry in this area may serve as a disease intermediary between the live bird market and the wild bird. Therefore, avoidance of the risk factors identified in this study, and implementation of preventive measures, may reduce the risk for infectious diseases. Governmental agencies need to develop prevention and control strategies, adapting to local requirements to reduce poultry disease loss as much as possible. These findings can be generalized and applied to similar poultry rearing environment in other developing countries, particularly in Southeast Asia.

## ACKNOWLEDGMENTS

We would like to thank the editor and anonymous reviewers for their insightful comments and constructive suggestions in improving the manuscript.

## REFERENCES

1 China Animal Industry Yearbook Editorial Committee (2011) China Animal Industry Yearbook. Beijing, China: Agriculture Press; 650pp (in Chinese).

2 National Development and Reform Commission (2006) National Animal Epidemic Prevention System Plan. Available: http://www.

ndrc.gov.cn/zcfb/zcfbtz/tz2006/W020070111591958303128.doc Accessed: 2013 March 20 (in Chinese).

3 Busani L, Valsecchi MG, Rossi E, Toson M, Ferrè N et al. (2009) Risk factors for highly pathogenic H7N1 avian influenza virus infection in poultry during the 1999-2000 epidemic in Italy. Vet J 181: 171-177.10.1016/j.tvjl.2008.02.013

4 Alexander DJ (2000) A review of avian influenza in different bird species. Vet Microbiol 74: 3-13.10.1016/S0378-1135(00)00160-7 PubMed: 10799774

5 Fouchier RAM, Munster V, Wallensten A, Bestebroer TM, Herfst S et al. (2005) Characterization of a novel influenza A virus hemagglutinin subtype (H16) obtained from black-headed gulls. J Virol 79: 2814-2822.10.1128/JVI.79.5.2814-2822.2005 PubMed: 15709000

6 Alexander DJ (2007) An overview of the epidemiology of avian influenza. Vaccine 25: 5637-5644.10.1016/ j.vaccine.2006.10. 051 PubMed: 17126960

7 Koch G, Elbers ARW (2006) Outdoor ranging of poultry: a major risk factor for the introduction and development of High-Pathogenicity Avian Influenza. Njas-Wagen. Life Sci 54: 179-194

8 Gilbert M, Chaitaweesub P, Parakarnawongsa T, Premashthira S, Tiensin T et al. (2006) Free-grazing ducks and highly pathogenic avian influenza, Thailand. Emerg Infect Dis 12: 227-234.10. 3201/eid1202.050640 PubMed: 16494747

9 Songserm T, Jam-on R, Sae-Heng N, Meemak N, Hulse-Post DJ et al. (2006) Domestic ducks and H5N1 influenza epidemic, Thailand. Emerg Infect Dis 12: 575-581

10 Prosser DJ, Cui P, Takekawa JY, Tang MJ, Hou YS et al. (2011) Wild Bird Migration across the Qinghai-Tibetan Plateau: A Transmission Route for Highly Pathogenic H5N1. PLOS ONE 6: e17622.10.1371/journal.pone.0017622 PubMed: 21408010

11 Liang L, Xu B, Chen YL, Liu Y, Cao WC et al. (2010) Combining Spatial-Temporal and Phylogenetic Analysis Approaches for

Improved Understanding on Global H5N1 Transmission. PLOS ONE 5: e13575.10.1371/journal.pone.0013575 PubMed: 21042591

12 Paul M, Wongnarkpet S, Gasqui P, Poolkhet C, Thongratsakul S et al. (2011) Risk factors for highly pathogenic avian influenza (HPAI) H5N1 infection in backyard chicken farms, Thailand. Acta Trop 118: 209-216.10.1016/j.actatropica.2011.03.009 PubMed: 21459074

13 Nishiguchi A, Kobayashi S, Yamamoto T, Ouchi Y, Sugizaki T et al. (2007) Risk factors for the introduction of avian influenza virus into commercial layer chicken farms during the outbreaks caused by a low-pathogenic H5N2 virus in Japan in 2005. Zoonoses Public Hlth 54: 337-343.10.1111/j.1863-2378.2007.01074.x PubMed: 18035971

14 Suarez DL, Senne DA (2000) Sequence analysis of related low-pathogenic and highly pathogenic H5N2 avian influenza isolates from United States live bird markets and poultry farms from 1983 to 1989. Avian Dis44: 356-364.10.2307/1592550 PubMed: 10879916

15 Kung NY, Morris RS, Perkins NR, Sims LD, Ellis TM et al. (2007) Risk for infection with highly pathogenic influenza A virus (H5N1) in chickens, Hong Kong, 2002. Emerg Infect Dis 13: 412-418.10.3201/eid1303.060365 PubMed: 17552094

16 Biswas PK, Christensen JP, Ahmed SSU, Das A, Rahman MH et al. (2009) Risk for Infection with Highly Pathogenic Avian Influenza Virus (H5N1) in Backyard Chickens, Bangladesh. Emerg Infect Dis 15: 1931-1936.10.3201/eid1512.090643 PubMed: 19961672

17 Akey BL (2003) Low-pathogenicity H7N2 avian influenza outbreak in Virginia during 2002. Avian Dis47: 1099-1103.10.1637/0005-2086-47.s3.1099 PubMed: 14575120

18 Mannelli A, Ferrè N, Marangon S (2006) Analysis of the 1999-2000 highly pathogenic avian influenza (H7N1) epidemic in the main poultry-production area in northern Italy. Prev Vet Med 73: 273-285.10.1016/j.prevetmed.2005.09.005 PubMed: 16243405

19 McQuiston JH, Garber LP, Porter-Spalding BA, Hahn JW, Pierson FW et al. (2005) Evaluation of risk factors for the spread of low

pathogenicity H7N2 avian influenza virus among commercial poultry farms.JAVMAJ AM Vetmed A 226: 767-772 PubMed: 15776951

20 Werner O, Starick E, Grund CH (2003) Isolation and characterization of a low-pathogenicity H7N7 influenza virus from a turkey in a small mixed free-range poultry flock in Germany. Avian Dis 47: 1104-1106.10.1637/0005-2086-47.s3.1104 PubMed: 14575121

21 Terregino C, De Nardi R, Guberti V, Scremin M, Raffini E et al. (2007) Active surveillance for avian influenza viruses in wild birds and backyard flocks in Northern Italy during 2004 to 2006. Avian Pathol 36: 337-344.10.1080/03079450701488345 PubMed: 17620182

22 Whiteford AM, Shere JA (2004) California experience with exotic Newcastle disease: a state and federal regulatory perspective. In: Proceedings of the 53rd Western Poultry. In. In: : Disease Conference, editor. pp 81-84

23 Shankman D, Liang Q (2003) Landscape Changes and Increasing Flood Frequency in China's Poyang Lake Region. Prof Geogr 55: 434-445.10.1111/0033-0124.5504003

24 Ji WT, Wang YB (2007) Survey on species and their number of wintering waterbirds in surrounding Poynag Lake area in 2006. China Crane. News 11: 37-38

25 Jiangxi Yearbook Editorial Committee (2011) Jiangxi Yearbook. Nanchang: Jiangxi People's Publishing House; 594pp (in Chinese).

26 Hulse-Post DJ, Sturm-Ramirez KM, Humberd J, Seiler P, Govorkova EA et al. (2005) Role of domestic ducks in the propagation and biological evolution of highly pathogenic H5N1 influenza viruses in Asia. Proc Natl Acad Sci U S A 102: 10682-10687.10.1073/pnas.0504662102 PubMed: 16030144

27 Sturm-Ramirez KM, Hulse-Post DJ, Govorkova EA, Humberd J, Seiler P et al. (2005) Are ducks contributing to the endemicity of

highly pathogenic H5N1 influenza virus in Asia? J Virol 79: 11269-11279.10.1128/JVI.79.17.11269-11279.2005 PubMed: 16103179

28 Fang LQ, de Vlas SJ, Liang S, Looman CWN, Gong P et al. (2008) Environmental factors contributing to the spread of H5N1 avian influenza in mainland China. PLOS ONE 3: e2268.10.1371/journal.pone.0002268 PubMed: 18509468

29 Paul M, Tavornpanich S, Abrial D, Gasqui P, Charras-Garrido M et al. (2010) Anthropogenic factors and the risk of highly pathogenic avian influenza H5N1: prospects from a spatial-based model. Vet Res 41: 28.10.1051/vetres/2009076 PubMed: 20003910

30 Sivanandan V, Halvorson DA, Laudert E, Senne DA, Kumar MC (1991) Isolation of H13N2 influenza-A virus from turkeys and surface-water. Avian Dis 35: 974-977.10.2307/1591638 PubMed: 1838479

31 Zhang HB, Xu B, Chen QJ, Chen JJ, Chen Z (2011) Characterization of an H10N8 influenza virus isolated from Dongting lake wetland. Virol J 8: 42.10.1186/1743-422X-8-42 PubMed: 21272297

32 Webster RG (2004) Wet markets - a continuing source of severe acute respiratory syndrome and influenza? Lancet 363: 234-236.10.1016/S0140-6736(03)15329-9 PubMed: 14738798

33 Guan Y, Shortridge KF, Krauss S, Chin PS, Dyrting KC et al. (2000) H9N2 influenza viruses possessing H5N1-like internal genomes continue to circulate in poultry in southeastern China. J Virol 74: 9372-9380.10.1128/JVI.74.20.9372-9380.2000 PubMed: 11000205

34 Liu M, He SQ, Walker D, Zhou NN, Perez DR et al. (2003) The influenza virus gene pool in a poultry market in South Central China. Virology 305: 267-275.10.1006/viro.2002.1762 PubMed: 12 573572

35 Hinshaw VS, Air GM, Gibbs AJ, Graves L, Prescott B et al. (1982) Antigenic and genetic-characterization of a novel hemagglutinin subtype of influenza-A viruses from gulls. J Virol 42: 865-872 PubMed: 7097861 [PMC free article]

36 Krauss S, Walker D, Pryor SP, Niles L, Li CH et al. (2004) Influenza A viruses of migrating wild aquatic birds in North America. Vectorborne Zoonot 4: 177-189.10.1089/ vbz.2004.4.177 PubMed:15631061

37 Li KS, Guan Y, Wang J, Smith GJD, Xu KM et al. (2004) Genesis Highly Pathogenic Potentially Pandemic H 5N1 influenza virus in eastern Asia. Nature 430: 209-213

38 Kou Z, Lei FM, Yu J, Fan ZJ, Yin ZH et al. (2005) New genotype of avian influenza H5N1 viruses isolated from tree sparrows in China. J Virol 79: 15460-15466.10.1128/JVI.79.24.15460-15466.2005 Pub Med: 16306617

39 Liu Q, Ma J, Kou Z, Pu J, Lei F et al. (2010) Characterization of a highly pathogenic avian influenza H5N1 clade 2.3.4 virus isolated from a tree sparrow. Virus Res 147: 25-29.10.1016/ j.virusres. 2009.09.014PubMed: 19804800

40 Chen H (2009) Avian influenza vaccination: the experience in China. Rev Sci Tech Off Int Epiz 28: 267-274 PubMed: 19618631

# CHAPTER 9

# Prevalence of *Campylobacter jejuni* in chicken produced by major poultry companies in Saudi Arabia

**Hany M Yehia[1,2], Mosffer M AL-Dagal[1]**

[1]Food Science and Nutrition Department, College of Food and Agricultural Sciences, King Saud University, Riyadh, Saudi Arabia.

[2]Food Science and Nutrition Department, Faculty of Home Economics, Helwan University, Cairo, Egypt.

## ABSTRACT

### Background

Campylobacter is a foodborne pathogen that is commonly associated with chicken. The aim of this work was to evaluate the prevalence of*Campylobacter jejuni* (as affected by refrigerated storage) in chicken samples obtained from the wholesale poultry market in the northern part of Riyadh City, Saudi Arabia.

## Findings

A gradual increase in the number of positive samples was noted during storage at 4°C. On days 1, 3, and 7, the number of positive samples were 10 (30.305%), 15 (45.45%), and 27 (81.81%), respectively. Of 99 tested samples, 52 (52.25%) were positive for *Campylobacter jejuni.* Protein profiling by Sodium dodecyl sulfate -Polyacrylamide gel electrophoresis (SDS-PAGE) was used to identify *Campylobacter jejuni.* The results were verified using Analytical Profile Index (API Campy system, Marcy l'Etoile, France). Forty-three (82.69%) positive isolates were identified as *C. jejuni subsp. jejuni* 2, 5 isolates as *C. jejuni subsp. jejuni* 1 (9.61%), and 4 isolates as *C. jejuni subsp.* doylei (7.69).

## Conclusion

*C. jejuni* positive samples increased rapidly during storage at 4°C for approximately 1 wk. Our results also indicated a connection between the protein profiles on SDS-PAGE and API Campy used for the identification of *C. jejuni.*

## KEYWORDS

Campylobacter jejuni Chicken Refrigeration storage

## 1. INTRODUCTION

*Campylobacter jejuni* is a well-known food-borne pathogen, transmitted to humans by the eating of warm-blooded animal meat, especially poultry (Jay et al. [2005]; Nielsen et al. [1997]). This bacterium is a major cause of food-borne diarrhea in many countries (Crushell et al. [2004]; Iovine et al. [2008]. It has gained more attention in the last 30 years because it has been recognized as a major cause of human illnesses, ranging from gastroenteritis to Guillain-Barré Syndrome (Khanna et al. [1996]; Tauxe, [2001]; Moore et al. [2005]). The 2 most frequently occurring *Campylobacter* species that are of clinical significance because of meat consumption and meat products are *C. jejuni* and*C. coli. Campylobacter jejuni* accounts for more than 90% of incidences of human campylobacteriosis (Lindmark et al. [2009]). Campylobacteriosis in humans results from eating undercooked meat and/or contaminated meals (Corry and Atabay [2001]). Survey studies have revealed a high prevalence of *Campylobacter* in poultry meats (Dickins et al. [2002]; Ridsdale et al. [1998]; Stoyanchev et al. [2007]).

Few studies on *Campylobacter* in the Saudi Arabian food market have been performed. Therefore, this study investigates the prevalence of*Campylobacter jejuni* in locally produced refrigerated chicken carcasses, as affected by storage time. Whole-cell protein profiles of presumptive Campylobacter isolates were compared with the standard strain of *Campylobacter jejuni* ATCC 33291 on SDS page. High degree of similarity within standard strain was confirmed by biochemical identification using an API CAMPY biotyping identification system.

# 2. MATERIALS AND METHODS

## 2.1 Sample collection

Whole chicken carcasses (n = 99) were obtained from a wholesale poultry market located in the northern part of Riyadh City, Saudi Arabia. The samples were collected from 11 major national poultry companies (designated by the letters A through K). Nine samples were collected from each company. The samples were transported at refrigeration temperature to the Food Microbiology Laboratory, College of Food and Agricultural Sciences, King Saud University.

## 2.2 Experimental design

Refrigerated chicken samples were divided into 3 groups. The first group (including 33 samples, 3 samples **(3 runs)** per company) was tested for the presence of *Campylobacter jejuni* on Day 1 (the purchase date). Similarly, the next 2 groups (33 samples each) were tested for the microorganism at Day 3 and Day 7 after the purchase date and stored in refrigerator at 4°C.

## 2.3 Isolation and identification of *Campylobacter*

*Campylobacter jejuni* was isolated according to the methods described by the Food Safety and Inspection Service (FSIS) ([1998]). Each carcass was rinsed in a sterile plastic bag with the addition of 200 mL of 0.1% peptone by manual shaking for 60 s. Ten milliliters of the rinse was centrifuged at 5000 rpm for 5 min, and 2 loops of the pellet were streaked on modified charcoal cefoperazone deoxycholate agar (mCCDA, Oxoid CM739, Basingstoke, Hampshire, UK) and *Campylobacter*-selective agar (Preston, Oxoid CM0689). Preston Campylobacter selective supplement (Oxoid, SR0117) and lysed horse blood (Oxoid, SR0048) were used for the selective isolation of*Campylobacter jejuni* from the samples (Figueroa et al. [2009]). As a confirmation step, 10 mL of each rinse fluid was transferred to 90 mL of Preston enrichment broth (to prepare Preston Campylobacter selective enrichment broth, 12.5 g of Nutrient Broth No.2 (Oxoid CM0067) was dissolved in 475 ml of distilled water and sterilized by autoclaving at 121°C for 15 minutes. Twenty five 25 ml of Lysed Horse Blood (Oxoid SR0048), 1 vial of Preston Campylobacter Selective Supplement (Oxoid SR0117) and 1 vial of Campylobacter Growth Supplement (Oxoid, SR0232) were added to the cooled 475 ml medium. Five ml volumes were aseptically dispensed in sterile small screw-capped bottles and incubated at 37°C for 48 h in a gas mixture of BBL GasPak, 70304 (Becton Dickinson and Cockeysville, MD, USA), (5% $O_2$, 10% $CO_2$, and 85% $N_2$). The enrichment was streaked onto selective media, and the plates were incubated at 42°C for up to 48 h under microaerophilic conditions. After incubation, the plates were inspected for presumptive colonies before Gram staining, and cells resembling with*Campylobacter* were subcultured onto mCCDA by streaking colony method and incubated for 2 to 5 d at 42°C under microaerophilic conditions.

## 2.4 Polyacrylamide Gel Electrophoresis (PAGE)

The preparation of isolates for SDS-PAGE and the running of the samples were performed according to the method by Scarcelli et al. [2001]. Electrophoresis was performed in a 12% polyacrylamide running gel and a 4% stacking gel, with a 0.025 M Tris 0.19 M glycine buffer pH 8.3, and 100 µL of a sucrose buffer (50 mM Tris–HCl, pH 8; 40 mM EDTA, pH 8; 0.75 M sucrose).

## 2.5 Preparation of cell extract

An overnight culture (100 µL) was inoculated into a 10 ml of fresh medium (Brain heart infusion-Oxoid, CM1135) and grown to an Optical Density (OD) 620 of 0.6 to 0.8 (3 to 4 h). The cells were collected and weighed, and 250 mg of cells were then suspended in 100 µL of a TES buffer (50 mM tris HCl, pH 8, 1 mM EDTA, 25% sucrose). Twenty microliters of lysozyme (50 mg/mL) and 5 µL mutanolysin (5000 u/mL) were added to the suspended cells in the TES buffer and incubated at 37°C for 30 min. Five to ten microliters of 20% SDS were added, and the contents were mixed until they became clear visible. The contents were stored at −20°C for 1 to 2 d (Ismail [2007]).

Fifty-microliter extracts (standard and isolated bacteria) were loaded on SDS-PAGE. Electrophoresis was performed at 25°C in a vertical tank apparatus using a constant-voltage power supply, until a bromophenol blue tracking dye reached the bottom of the gel. Gels were stained with 0.25% Coomassie Brilliant Blue R-250 (Bio-Rad, Marnes-la-Coquette, France) in water: methanol: acetic acid (6.5:2.5:1) for 18 h at room temperature. Gel destaining was performed by continuous agitation in a methanol: acetic acid: water (20:10:70 v/v/v) solvent until obvious bands of proteins were obtained.

Whole-cell protein profiles of presumptive *Campylobacter* isolates were compared with the standard strain of *Campylobacter jejuni* ATCC 33291 on SDS-PAGE. A high degree of similarity with the standard strains was confirmed by biochemical identification using an API CAMPY biotyping system and a catalase test.

## 2.6 Biochemical identification of *Campylobacter* isolates

BioMérieux API CAMPY was used according to the manufacturer's instructions for the biochemical identification of *Campylobacter*. The BioMérieux API CAMPY strip consisted of 20 microtubes containing dehydrated substances, with each microtubule corresponding to an individual test. The 20 tests were divided into 2 parts; the first part was composed of enzymatic and conventional tests, and the second part comprised assimilation or inhibition tests (Gorman and Adley [2005]). The results of the enzymatic tests were obtained with the addition of conventional reagents after 24 h of incubation at 37°C under aerobic conditions. The results of the assimilation and inhibition tests were recorded after 24 h at 37°C under microaerophilic conditions. Incubation was extended to 48 h if the succinate assimilation test was negative, as indicated by the manufacturer's instructions.

# 3. RESULTS AND DISCUSSION

Table 1 shows the prevalence of *C. jejuni* in refrigerated chicken samples, as affected by storage time. According to the table, the samples from all 11 poultry companies contained *C. jejuni* at one or more of the tested shelf-life dates. At Day 1 of the shelf life, *C. jejuni* was recovered from 30.30% of the tested samples (10 of 33 samples). This percentage increased to 44.45% and to 81.81%, after refrigeration for 3 d and 7 d, respectively, indicating the positive effect of natural enrichment. Poultry companies B, C, I, and J had included highly positive samples for *C. jejuni* (n = 6; 66.7%), followed by A and E (n = 5; 55.6%), and then D, F, and K (n = 4; 44.4%). Only one sample (n = 1; 11.1%), from company G, was positive.

**Table 1** Prevalence of ***Campylobacter jejuni*** in refrigerated whole chicken carcasses of 11 poultry companies in Riyadh City wholesale market

| Poultry samples | Group 1 First day of the production date | | | Group 2 $3^{rd}$days of the production date | | | Group 3 $7^{th}$ day of the production date | | | Total Positive (out of 9) | % |
|---|---|---|---|---|---|---|---|---|---|---|---|
| | Run1 | Run2 | Run3 | Run1 | Run2 | Run3 | Run1 | Run2 | Run3 | | |
| A | + | - | - | + | - | - | + | + | + | 5 | 55.6 |
| B | - | - | + | + | - | + | + | + | + | 6 | 66.7 |
| C | - | + | - | + | + | + | + | + | - | 6 | 66.7 |
| D | - | + | - | - | - | - | + | + | + | 4 | 44.4 |
| E | + | - | - | + | - | + | - | + | + | 5 | 55.6 |
| F | - | - | + | + | - | + | - | + | - | 4 | 44.4 |
| G | - | - | - | - | - | - | - | + | - | 1 | 11.1 |
| H | + | - | - | + | - | - | + | + | + | 5 | 55.6 |
| I | - | + | - | + | - | + | + | + | + | 6 | 66.7 |
| J | - | - | + | + | | + | + | + | + | 6 | 66.7 |
| K | - | + | - | - | - | - | + | + | + | 4 | 44.4 |
| Number of total samples | 33 | | | 33 | | | 33 | | | 99 | |
| Number of positive sample | 10 | | | 15 | | | 27 | | | 52 | |
| % | 30.30 | | | 45.45 | | | 81.81 | | | 52.52 | |

Survival of *C. jejuni* at refrigeration temperature approved in one study. The outcome of this work agreed with other research works indicating that the survival of *C. jejuni* at refrigeration temperature range (4 to 7°C) is better than range (20 to 30°C), and that it exhibits greater survivability at chilled temperatures (Karenlampi and Hanninen [2004]; El-Shibiny et al. [2009]). Rollins and Colwell ([1986]) showed that, at 4°C, *C. jejuni* can survive and remain at a viable but non-culturable stage for approximately 4 mo. Zhao et al. ([2000]) showed that *C. jejuni*survive for days or weeks in refrigerated foodstuffs. Vashin and Stoyanchev ([2011]) established that the microorganisms did not grow in chilled or frozen meat, but are able to survive during the storage period at 1-4°C. Campylobacter were detected up to the 25$^{th}$ days while at -18 to −20°C: up to the 45$^{th}$ day. The juice released into the bags from poultry liquidation is highly nutritive and forms microaerophilic conditions suitable for *Campylobacter*. Birk et al. ([2004]) confirmed that compounds present in chicken juice protect, and thus, prolong the survival of *C. jejuni* during storage at refrigerator temperatures. They confirmed that incubation at 5°C extended viability of cells of *C. jejuni*, and incubation at 10°C significantly prolonged the viability of *Campylobacter*. In addition, they found that storage in chicken juice at both 5°C and 10°C significantly prolonged the viable cells of *C. jejuni* compared to incubation in reference media. The total number of samples tested positive for *C. jejuni* was 52 (52.52%). Similar results in a previous survey study (Rahimi and Tajbakhsh [2008]) found 56.1% of chicken samples to be positive for *Campylobacter*. Few studies related to the occurrence of Campylobacter in Saudi Arabian food market, from 2002 to 2004, two studies were conducted to assess specimens obtained from slaughter-houses in Bahrain and Saudi Arabia for *Campylobacter*contamination. In one study, specimens consisting of 35 whole chickens, 27 chicken livers, and 38 chicken faeces were assessed using a combination of three culture methods, and just over half (57%) were found to be positive for *Campylobacter* contamination (Ghazwan [2006]). In another study, 60 chicken faeces specimens were assessed using a newly developed multiplex PCR technique with 100% *Campylobacter*detection (Al Amri et al. [2007]). However, in both studies, *C. jejuni* accounted for the majority of *Campylobacter* detected. The findings of this study, which is the first of its kind in our setting, indicates a need for increased surveillance and *Campylobacter* screening in food safety control to better protect consumers.

Figure 1 shows the protein profiles of the isolates as they appeared on the media of mCCDA and Preston agar compared with standard strains of *C. jejuni* ATCC 33291 on SDS-PAGE. The results demonstrated the presence of the common heavy protein band for 52 isolates compared with *C. jejuni* ATCC 33291. The whole-cell protein profiles of isolates were closely related to *C. jejuni* ATCC 33291. Some isolates could not be analyzed because they did not yield sufficient proteins after extraction, or the band did not resemble *C. jejuni* ATCC 33291. Any isolates showing discrepancy with the protein profile were excluded from confirmation of identification by using API CAMPY. Massai et al. ([2007]) explained that the whole-cell protein profile determined by SDS-PAGE expresses an important proportion of the genome. However, this expression may be modified by various factors. Therefore, this technique must be carefully controlled and standardized to obtain reproducible results. Advantages of the SDS-PAGE patterns are that they can be obtained in a short time, are reproducible, and do not require any sophisticated and expensive reagents or equipment compared with other molecular biology.

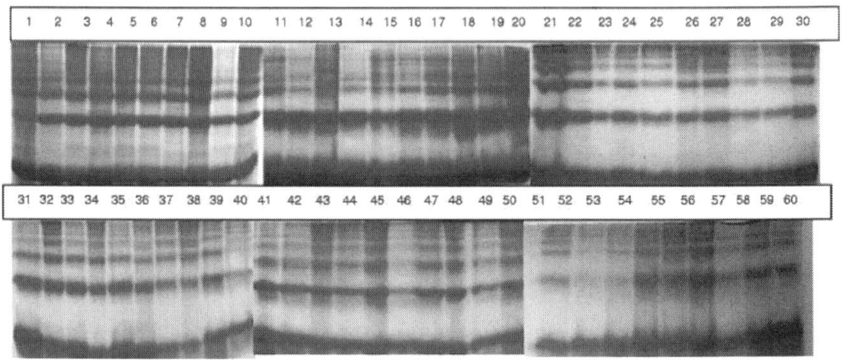


**Figure 1** Comparison of patterns from of whole-cell protein profile of references strain ***C. jejuni*** ATCC 33291 (lane 1), and isolates from poultry carcass (lanes 2–60) on 12% SDS-PAGE.

The polyacrylamide gel electrophoresis (PAGE) of bacterial proteins is an efficient technique for the classification of microorganisms, based on phenotypical characteristics expressed by their protein profiles (Scarcelli et al. [2001]). Protein electrophoretic analysis of *Campylobacter spp.* in a polyacrylamide gel in the presence of SDS supplies data that can be used in epidemiological and taxonomic studies, as well as the identification of species and specific virulence factors (Dunn et al. [1987]).

## 3.1 API campy strip system

After 24 to 48 h of incubation, the positive isolates obtained from SDS-PAGE were confirmed using API Campy, with 52 of 99 samples (52.52%) testing positive for both the first part and second part of the tests. Eleven of 99 (11.11%) isolates were negative. Forty-three of the 52 (82.69%) positive isolates were identified in both parts of the tests as *C. jejuni subsp. jejuni* 2. Five isolates were identified as *C. jejuni subsp. jejuni* 1 (9.61%), and 4 isolates were identified as *Campylobacter jejuni subsp. doylei* (7.96%). The reference strains were correctly identified by both systems as *C. jejuni ssp. jejuni* 1. Eleven isolates yielded discrepant identifications, and yielded profile codes absent in the API Campy database; therefore, they were considered negative results. Huysmans and Turnidge ([1997]) indicated that the correlation between API Campy and the conventional tests was 100% for the identification of *C. jejuni*.

# 4. CONCLUSION

The results show that *Campylobacter jejuni* was prevalent in poultry meat samples collected from wholesale markets in Riyadh City, Saudi Arabia. *C. jejuni* positive samples increased rapidly during storage at 4°C for approximately 1 wk. Our results also indicated a connection between the protein profiles on SDS-PAGE and API Campy used for the identification of *C. jejuni*.

## COMPETING INTERESTS

The authors declare that they have no competing interests.

## AUTHORS' CONTRIBUTIONS

Conceived and designed the experiments: YHM and AMM. Performed the experiments: YHM . Analyzed the data: YHM and AMM. Wrote the manuscript: YHM and AMM. All authors read and approve the final manuscript.

## REFERENCES

1 Al Amri A, Senok AC, Ismaeel AY, Al-Mahmeed AE, Botta GA: Multiplex PCR for direct identification of *Campylobacter* spp. in human and chicken stools. *J Med Microbiol* 2007, 56: 1350–1355. 10.1099/jmm.0.47220-0

2 Birk T, Ingmer H, Andersen MT, Jorgensen K, Brondsted L: Chicken juice, a food-based model system suitable to study survival of*Campylobacter jejuni* . *Lett Appl Microbiol* 2004, 38: 66–71. 10.1046/j.1472-765X.2003.01446.x

3 Corry JE, Atabay HI: Poultry as a source of Campylobacter and related organisms. *Symp Ser Soc Appl Microbiol* 2001, 30: 96S-114S. 10.1046/j.1365-2672.2001.01358.x

4 Crushell E, Harty S, Sharif F, Bourke B: Enteric Campylobacter: purging its secrets? *Pediatr Res* 2004, 55: 3–12. 10.1203/01. PDR. 0000099794.06260.71

5 Dickins MA, Franklin S, Stefanova R, Schutze GE, Eisenach KD, Wesley I, Cave MD: Diversity of Campylobacter isolates from retail poultry carcasses and from humans as demonstrated by pulsed-field gel electrophoresis. *J Food Prot* 2002, 65: 957–962.

6 Dunn BE, Blaser MJ, Snyder EL: Two dimensional gel electrophoresis and immunoblotting of *Campylobacter* outer membrane proteins. *Infect Immun* 1987, 55: 1564–1572.

7 El-Shibiny A, Scott A, Timms A, Metawea Y, Connerton P, Connerton I: Application of a group II Campylobacter bacteriophage to reduce strains of *Campylobacter jejuni* and *Campylobacter coli* colonizing broiler chickens. *J Food Prot* 2009, 72: 733–740.

8 Figueroa G, Troncoso M, Lopez C, Rivas P, Toro M: Occurrence and enumeration of *Campylobacter spp* . during the processing of Chilean broilers. *BMC Microbiol* 2009, 9: 94. 10.1186/1471-2180-9-94

9 *Isolation, Identification and Enumeration of Campylobacter jejuni/coli from meat and poultry products*. Microbiology Laboratory Guidebook. Chapter 3rd edition, United States Department of Agriculture, Washington D.C.; 1998.

10 Ghazwan J: *Public Health Importance of Campylobacter jejuni in Poultry [dissertation]*. Arabian Gulf University, Manama, Bahrain; 2006.

11 Gorman R, Adley CC: Campylobacter Isolation, Identification, and Preservation .From: Methods in Biotechnology, Vol. 21. In *Food Borne Pathogens: Methods and Protocols*. Edited by: Adley CC. © Humana Press Inc, Totowa, NJ; 2005.

12 Huysmans MB, Turnidge JD: Disc susceptibility testing for thermophilic Campylobacters. *Pathology* 1997, 29: 209–216. 10. 1080/00313029700169884

13 Iovine NM, Pursnani S, Voldman A, Wasserman G, Blaser MJ, Weinrauch Y: Reactive nitrogen species contribute to innate host defense against *Campylobacter jejuni* . *Infect Immun* 2008, 76: 986–993. 10.1128/IAI.01063-07

14 Ismail EA: *Ph. Thesis Characterization and Genetic Improvement of Lactobacillus for Application in Probiotics Dairy Probiotic Dairy Products*. University of Kiel, Germany; 2007.

15 Jay J, Loessner M, Golden DA: *Modern Food Microbiology*. Springer, New York; 2005.

16 Karenlampi R, Hanninen ML: Survival of *Campylobacter jejuni* on various fresh produce. *Int J Food Microbiol* 2004, 97: 187–195. 10.1016/j.ijfoodmicro.2004.04.019

17 Khanna PN, Kumar A, Singh AK, Khan IA: Thermophilic Campylobacters - public health importance and our observations. *Indian J Comp Microbiol Immunol Infect Dis* 1996, 17: 32–40.

18 Lindmark H, Boqvist S, Ljungstrom M, Agren P, Bjorkholm B, Engstrand L: Risk factors for Campylobacteriosis: an epidemiological surveillance study of patients and retail poultry. *J Clin Microbiol* 2009,47(8):2616–2619. 10.1128/JCM.00826-09

19 Massai R, Bantarc C, Lopardo H, Vay C, Gutkind G: Whole-cell protein profiles are useful for distinguishing enterococcal species recovered from clinical specimens. *Rev Argent Microbiol* 2007, 39: 199–203.

20 Moore JE, Corcoran D, Dooley JSG, Fanning S, Lucey B, Matsuda M, McDowell R, O'Riordan L, O'Rourke M, Rao JR, Rooney PJ, Sails A, Whyte P: Campylobacter. *Vet Res* 2005, 36: 351–382. 10.1051/vetres:2005012

21 Nielsen EM, Engberg J, Madsen M: Distribution of serotypes of *Campylobacter jejuni* and *C. coli* from Danish patients, poultry, cattle and swine. *FEMS Immunol Med Microbiol* 1997, 19: 47–56. 10.1016/S0928-8244(97)00049-7

22 Rahimi E, Tajbakhsh E: Prevalence of Campylobacter species in poultry meat in the Esfahan city, Iran. *Bulg J Vet Med* 2008, 11:257–262.

23 Ridsdale JA, Atabay HI, Corry JEL: Prevalence of Campylobacters and Arcobacters in ducks at the abattoir. *J Appl Microbiol* 1998,85: 567–573. 10.1046/j.1365-2672.1998.853537.x

24 Rollins DM, Colwell RR: Viable but non culturable stage of *Campylobacter jejuni* and its role in survival in the natural aquatic environment. *Appl Environ Microbiol* 1986, 52: 531–538.

25 Scarcelli E, Elizabeth OC, Margareth EG, Maristela VC, Erna EB, Ana PT: Comparison of electrophoretic protein profiles of*Campylobacter jejuni subsp. Jejuni* isolated from different animal species. *Braz J Microbiol* 2001,32(4):286–292. 10.1590/S1517-83822001000400006

26 Stoyanchev T, Vashin I, Ring C, Atanassova V: Prevalence of *Campylobacter spp* . in poultry and poultry products for sale on the Bulgarian retail market. *Anton Leeuw* 2007, 92: 285–288. 10.1007/s10482-007-9154-6

27 Tauxe RV: *Incidence, Trends and Sources of Campylobacteriosis in Developed Countries: An Overview. In the Increasing Incidence of Human Campylobacteriosis. Report and Proceedings of a WHO Constitution of Experts Copenhagen, Denmark, 21–25 November 2000, pp. 42–43*. World Health Organization, Geneva; 2001.

28 Vashin IT, Stoyanchev TT: Influence of temperature on *Campylobacter jejuni* survival rates in pork meat. *Bulg J Vet Med*2011,14(1):25–30.

29 Zhao T, Doyle MP, Berg DE: Fate of *Campylobacter jejuni* in butter. *J Food Prot* 2000, 63: 120–122.

# CHAPTER 10

# An infected chicken kidney cell co-culture ELISpot for enhanced detection of T cell responses to avian influenza and vaccination

**Raul Ruiz-Hernandez[a], Marylene Peroval[a], Amy Boyd[b], Devanand Balkissoon[a], Karen Staines[a], Adrian Smith[b], Colin Butter[a,*]**

[a] The Pirbright Institute, Compton Laboratory, Compton, Newbury, Berkshire RG20 7NN, UK
[b] University of Oxford, Department of Zoology, South Parks Road, Oxford OX1 3PS, UK

## ABSTRACT

A better understanding of the immune responses of chickens to the influenza virus is essential for the development of new strategies of vaccination and control. We have developed a method incorporating infected chicken kidney cells (CKC) in culture with splenocytes in an IFNγ ELISpot assay to enumerate ex vivo responses against influenza virus antigens. Splenocytes from birds challenged with influenza showed specific responses to the influenza virus, with responding cells being mainly CD8 positive. The utility of the assay was also demonstrated in the detection of an antigen specific enhancement of IFNγ producing cells from birds vaccinated with recombinant Fowlpox vectored influenza nucleoprotein and matrix protein.

## KEYWORDS

Adaptive antiviral responses; Antigen presentation; Avian influenza; IFNγ ELISpot; CD8 T cell

## 1. INTRODUCTION

Avian influenza viruses (AIVs) belong to the Orthomyxoviridae family and are classified according to their haemagglutinin (HA) and neuraminidase (NA) proteins. On the basis of their ability to cause disease in poultry, avian influenza viruses are further classified as low pathogenic (LPAI) and highly pathogenic (HPAI), both causing severe financial losses to the poultry industry. Poultry also act as a reservoir for AIVs and thus provide an environment for the emergence of novel AIV subtypes, which may present a threat to human health, through the processes of recombination and re-assortment. Hence, improved understanding of influenza virus infections in chickens is an important aspect of developing new control measures, including vaccines for use in poultry. Improved control of influenza in chickens will protect the poultry industry and reduce the risk of zoonotic transfer to humans.

Although influenza viruses are endemic in avian species (Stech et al., 2009) understanding of influenza-specific cellular responses is more limited in chickens than in humans or mice; until recently a paucity of reagents and techniques has impeded a comprehensive study in birds. Although most studies of host responses to influenza infection or vaccination in birds have focused on the production of neutralizing antibodies, it is clear that cell mediated immunity (CMI) is also relevant (Suarez and Schultz-Cherry, 2000). The principal route for the presentation of viral antigenic peptides involves the major histocompatibility complex I (MHC I) pathway and results in antigen presentation to CD8⁺ T cells (Subbarao and Joseph, 2007). In birds and mammals, influenza-specific CD8⁺ cytotoxic cells become activated and produce IFNγ during infection in response to the engagement of their T cell receptors with influenza-derived peptides in the context of MHCI on the surface of antigen presenting cells (APC) (Rock et al., 1990, Suarez and Schultz-Cherry, 2000, Novak et al., 2001 and Subbarao and Joseph, 2007). Cytotoxic T cell responses can be generated against a variety of influenza proteins including surface associated HA and NA antigens as well as internal proteins such as matrix protein (M1) and nucleoprotein (NP). These CD8⁺ T cell responses contribute to the control of influenza virus replication within cells, thereby enabling viral clearance and limiting viral spread (Suarez and Schultz-Cherry, 2000 and Kwon et al., 2008). A suppression of these responses may contribute to high and disseminated viral replication in chickens, contributing to the pathogenicity of LPAI viruses (Kwon et al., 2008). Enhancement of CTL responses (principally against the most conserved epitopes) has thus become the aim of vaccines designed to produce heterosubtypic immunity to influenza in humans (Berthoud et al., 2011) and chickens (Boyd et al., 2012).

A number of techniques and reagents are currently in use for the study of T cell responses in poultry. These include the measurement of antigen specific proliferation by flow cytometry (Dalgaard et al., 2010) intracellular cytokine staining (De Boever et al., 2010), measurement of IFNγ production by Enzyme-Linked Immunosorbent Assay (ELISA) (Ariaans et al., 2009 and Rauw et al., 2011) and Enzyme-Linked Immunosorbent Spot (ELISpot) assay (Ariaans et al., 2008 and Ariaans et al., 2009). CTL responses to infectious bronchitis virus have previously been monitored using MHC matched chicken kidney cells (CKC) serving as antigen presenting cells (APC) (Seo and Collisson, 1997). Through the use of MHC matched infected cells as surrogate APC, the measurement of chicken IFNγ responses against whole influenza virus or viral proteins has been achieved using an indirect method based on the ability of IFNγ to activate the HD11 macrophage cell line (Gobel et al., 2003, Singh et al., 2010a and Singh et al., 2010b).

Peptides are often used to study antigen specific responses, and this method has been applied successfully in birds (Haghighi et al.,

2009 and Reemers et al., 2012). While the use of peptide libraries to identify influenza antigen specific responses can be exquisitely informative, it also has his limitations. The cost of peptide libraries can be prohibitive for many labs, even before technical considerations. The use of a library of predicted binding peptides excludes epitopes that are not predicted due to an incomplete understanding of the binding motifs of chicken class I MHC. This may be particularly challenging with haplotypes such as B21 that has a highly promiscuous motif (Koch et al., 2007). Although peptide length and motif can give an indication of which group of cells responds, this does not provide definitive information regarding the phenotypic identification of effector T cells (CD4 versus CD8), and there are no significant data regarding processes such as cross presentation in poultry model, rendering interpretation of peptide data difficult. In addition, techniques such as intracellular staining are technically challenging, requiring many manipulations of the cells. ELISpot, while sensitive, provides no information as to the effector cell phenotype unless the responding cells are sorted prior to plating. This is also true for the HD11 activation method, which requires culture and stimulation of HD11 as an extra step.

In the present study we set out to develop a method to preferentially detect CD8 T cell responses. We hypothesized that by infecting cells which only express class I MHC with AIVs and culturing these with splenocytes from infected birds we would potentiate detection of influenza specific $CD8^+$ T cells. We first derived a chicken kidney cell line from inbred chickens MHC matched to our experimental cohort, tested that the CKC expressed MHCI but not MHCII, then infected them with influenza virus. After irradiation these cells were co-cultured in ELISpots with MHC matched splenocytes either from chickens exposed to influenza virus or from vaccinated chickens. In both experimental scenarios we were able to demonstrate the presence of antigen specific T cells. We also demonstrated by flow cytometry that the IFNγ producing cells were principally CD8 positive. The assay was reproducible, with high sensitivity and low background noise, and will be a useful tool in the analysis of CD8 T cell responses.

# 2. MATERIAL AND METHODS

## 2.1. Animals

Inbred lines of White Leghorn chickens, Line O (haplotype B21) or Line 15 (B15) (Miller et al., 2004), were produced and maintained at the Pirbright Institute (Compton, UK) in specific pathogen-free (SPF) conditions and fed ad libitum. For infection studies birds were housed in self-contained BioFlex® B50 Rigid Body Poultry isolators (Bell Isolation Systems). Animal procedures were carried out in accordance with local ethical review and UK Home Office requirements (Home Office, 1986).

## 2.2. Viral stocks

LPAI virus (A/Turkey/England/1977/H7N7) was grown in embryonated chicken eggs using standard methods described elsewhere (World Health Organization. Dept. of Epidemic and Pandemic Alert and Response., 2002). Viral titer was estimated by plaque assay on Madin-Darby canine kidney (MDCK) cells, using standard techniques (Gaush and Smith, 1968).

Virus was inactivated in a final concentration of 0.094% β-propiolactone (ACROS Organics, Geel, Belgium), as described previously (Jonges et al., 2010) and aliquots were stored at − 80 °C until its use. Inactivation was verified by the absence of plaques on MDCK cells. Recombinant Fowlpox virus (rFPV) vectors expressing NP and M1 transgenes from avian influenza A/Turkey/Turkey/1/2005 (H5N1) or GFP were the kind gift of Dr. Mike Skinner (Imperial College). Modified Vaccinia Ankara (MVA) virus expressing a fusion protein of nucleoprotein and matrix protein 1 (MVA-NpM1) from influenza A/Panama/2007/99 (H3N2) was supplied by the Vector Core Facility at the Jenner Institute (Oxford, UK) (Berthoud et al., 2011).

## 2.3. Vaccination and infection

In a first round of experiments, 3 week old birds were randomly allocated to infected or control groups. Birds were challenged by intranasal inoculation of LPAI (A/Turkey/England/1977 H7N7) at a dose of $3.4\times10^{7}$ pfu in 100 µl PBS per bird. In the second round of experiments, birds were vaccinated subcutaneously with $10^{5}$ pfu rFPV at 1 day old, boosted with the same dose at 9 days old, and challenged with LPAI, as above, at 4 weeks old. Birds were killed 10 days post-infection.

## 2.4. Measurement of viral shedding

Sterile polyester tipped swabs (Fisher Scientific, UK) were used to sample buccal cavities, transferred to a solution of viral transport media (World Health Organization. Dept. of Epidemic and Pandemic Alert and Response., 2002), vortexed briefly, clarified of debris by centrifugation at 450 ×*g* for 2 min and stored at –80 °C. RNA was extracted using the QIAamp One-For-All Nucleic Acid Kit on an automated extraction platform (Qiagen BioRobot Universal System, Qiagen, UK) according to the manufacturer's instructions. Quantitation of influenza virus in RNA from swabs was performed by analysis of matrix gene transcripts. A single step real-time reverse transcriptase PCR was carried out using the Superscript III Platinum One-Step qRT-PCR Kit (Life Technologies, UK). Primers and a probe specific for a conserved region of the Influenza A Matrix gene were used as described previously (Spackman et al., 2002). Cycling conditions were: 50 °C, 5 min; 95 °C, 2 min; and then 40 cycles of 95 °C, 3 s and 60 °C, 30 s, using a 7500 fast real-time PCR machine (Applied Biosystems, UK). Results are expressed in terms of the threshold cycle value (Ct), the cycle at which the change in the reporter dye signal passes a significance threshold (Rn).

## 2.5. Cell lines

MDCK cells were grown in Dulbecco's Modified Eagles Medium (DMEM) with Glutamax (Life Technologies), supplemented with non-essential Amino Acids (Sigma), 100 U/ml penicillin, 100 µg/ml streptomycin and 10% fetal bovine serum (FCS). Chinese hamster ovary (CHO) cells were grown in Ham's F12 medium (Life Technologies) with 10% FCS. Puromycin HCl (Enzo) was used at 20 μg/ml for selection of IFNγ transfected lines and at 15 μg/ml for maintenance of transfected CHO cells. Cell cultures were maintained in 5% $CO_2$ at 37 °C.

Primary chicken kidney cell (CKC) lines were established from 10 day old birds following guidelines previously described (Seo and Collisson, 1997). Briefly, cells were dispersed with trypsin digestion and cultured in 150 or 75 $cm^2$ tissue culture flasks. The CKC adherent cells were continuously cultured by passage every 4–6 days in Minimum Essential Medium (MEM) supplemented with tryptose phosphate broth (TPB), glutamine, 1M HEPES, fungizone, 100 U/ml penicillin, 100 μg/ml streptomycin and 10% FCS. Chicken cell cultures were maintained in 5% $CO_2$ at 41 °C.

## 2.6. Generation of anti-chicken IFNγ monoclonal antibodies

Antibodies were generated using a technique previously described (Staines et al., 2013). Briefly, chicken IFNγ was amplified from a spleen cDNA library using the following primers; IFN-Foward-NheI (5′-AGCCATCAGCTAGCAGATGACTTG) and IFN-Reverse-BglII (5′-ATCTCCTCAGATCTTGGCTCCTTTTC) and cloned into an Ig-fusion protein vector. To obtain ChIFNγ monoclonal antibodies, we immunized mice with two intramuscular injections of 100 μg of the IFNγ-IgG1Fc plasmid diluted in PBS (endotoxin free, Qiagen Endofree Plasmid Maxi Kit) at four week intervals. After a further four weeks, mice received a final boost with an intraperitoneal injection of 50 μg purified fusion protein and were sacrificed four days later for preparation of splenocytes which were fused with NS0 hybridoma partner cells using established methods. Hybridoma supernatants were first screened by ELISA for antibodies binding fusion protein immobilized with anti-human IgG and detected with HRP conjugated goat anti-mouse IgG. Antibodies recognizing the human Ig moiety of the fusion protein were eliminated by a similar ELISA using a control fusion protein containing the same human IgG1 sequence (DEC205-IgG1Fc, (Staines et al., 2013)).

Antibodies from two IFNγ-specific clones, AF10 and EH9, were purified from high density culture (miniPERM, Sarstedt) with Hi Trap Protein G HP columns (Amersham-Pharmacia, UK) according to the manufacturer's instructions. After dialysis against PBS, the concentration of these antibodies was estimated by measurement of the absorbance at 280 nm.

## 2.7. Preparation of infected CKC

CKC were infected with A/Turkey/England/1977/H7N7 for use in co-culture as previously described (Singh et al., 2010a). Briefly, confluent monolayers of CKC (after a minimum of 8 passages) were infected with AIVs for 1 h at a Multiplicity of Infection (MOI) of 3–5, washed with PBS, and incubated for 4 h with CKC growth media without FCS, supplemented with TPCK trypsin (Sigma). Cells were then washed, dispersed with trypsin, washed again, counted, resuspended in leukocyte culture media and then irradiated with 3000 rad using a Gammacell 1000 Elite caesium 137 gamma irradiator (Nordion, Canada).

For infection with recombinant MVA, CKCs were infected by incubation for 1 h at 37 °C at an MOI of 5. We optimized these conditions through analysis of GFP transgene expression by confocal microscopy (Supplementary Fig. 1). Following incubation, cells were washed, counted, irradiated as described, and resuspended in leukocyte culture media. The irradiated CKC were used at a ratio of 1:10 (CKC:splenocyte) in co-culture ELISpot.

For confocal imaging $5\times10^4$ primary CKC in growth media per chamber of an 8 chamber slide (Lab-TekII, Nunc) were incubated at 41 °C, 5% $CO_2$, for 1 day. Any non-adherent cells were discarded and the adherent cell population was infected with MVA-GFP constructs as described above. After incubation, cells were fixed with a solution of 4% paraformaldehyde for 20 min, and then washed in PBS. Nuclei were stained by incubation with 2 μg/ml DAPI (Sigma) for 10 min. Sections were mounted in Vectashield (Vector Laboratories) and analyzed using a confocal microscope (Leica SP2 with 405-, 488-, and 568-nm lasers).

## 2.8. IFNγ ELISpot

Spleens were macerated in cold sterile PBS and passed through a 100 μm cell strainer (Fisher, UK). Cell suspensions were centrifuged at 220 × *g* for 10 min at 4 °C and resuspended in culture media (RPMI 1640 medium with Glutamax supplemented with 10% FCS, 100 U/ml penicillin, and 100 μg/ml streptomycin) (all from Life Technologies, UK) before under-laying Histopaque 1119 (Sigma, UK) and centrifuged at 2000 rpm (492 ×g) for 20 min at 4 °C. Cells harvested from the interphase were washed twice, counted using a Countess™ automated cell counter (Life Technologies) and resuspended at 5 × $10^6$/ml. ChIFNγ ELISpot was carried out as described previously (Ariaans et al., 2008), using either antibodies from a commercially available kit for detection of chicken IFNγ protein (chicken IFNγ ELISA kit, Life Technologies ®) or EH9/AF10 antibodies produced as described. 96-well plates (MAIPS4510, Millipore) were coated overnight at 4 °C with the commercial capture antibody or EH9, diluted in pH 9.6 50 mM carbonate/bicarbonate buffer at a final concentration of 2 μg/ml, and washed with PBS + 0.1% Tween 20 at this stage and between all subsequent steps. Plates were blocked with experiment culture media. Chicken leukocyte suspensions consisting of 3 or 5 × $10^5$ cell/well were maintained in 5% $CO_2$ at 41 °C for 1 or 2 days. Cells were incubated in the presence of either culture medium or medium supplemented with one of the following stimuli to a final volume of 200 μl per well: phorbol 12-myristate 13-acetate (PMA 500 ng/ml) plus ionomycin (750 ng/ml, Sigma); Concanavalin A (ConA, 10 μg/ml final; Sigma); inactivated or live virus (MOI 3-5); prepared exogenous APCs (1:10, CKC: splenocyte), pooled or individual peptide (5 μM). A library of 62 overlapping peptides spanning NP protein from A/Turkey/England/1977/H7N7 virus (challenge virus) was synthesized commercially (Neobioscience, Massachusetts, USA), resuspended in DMSO or in a solution of 50% acetic acid in water, aliquots stored at − 20 °C until use, and then diluted to a final concentration of 5 μM in culture wells. Peptides were 18 aa long and 10 aa overlapping (Supplementary Table 1). When the chicken IFNγ ELISA kit (Life Technologies®) was used, ELISpot plates were then incubated at RT with the biotinylated detection antibody (1 μg/ml) followed by an incubation with streptavidin-horseradish peroxidase conjugate at a

1/2000 dilution. Otherwise plates were incubated with the detection antibody AF10, followed by an incubation with first a biotinylated goat anti-mouse anti isotype IgG2b (AF10 isotype) antibody (Southern biotech) followed by avidin-HRP (Southern Biotech). Plates were developed by incubation with 100 µl per well of 3-amino-9-ethylcarbazole, (AEC, Merck Chemicals, UK). After spot development, plates were rinsed with tap water and allowed to dry overnight before counting using an ELISpot plate reader (AID systems, Germany). Results were expressed as number of spots (SFU, spot forming unit) per $10^6$ splenocytes. Depending on the stimuli used, experiments were carried out in triplicate (whole virus or CKCs) or in duplicate (peptides).

## 2.9. Serum sample processing and hemagglutination inhibition assays

Blood samples (0.5–1 ml/bird) from all challenged birds were drawn from a wing vein 2 weeks after infection to evaluate humoral responses against influenza virus. These were left to clot at room temperature (RT) and sera were retrieved after centrifugation and stored at − 20 °C until analysis. A standard HI test was used to measure serum AIV antibody titers, which were expressed in log2 mean HI titers in each sample for each group (Spackman, 2008).

## 2.10. Flow cytometry and intracellular staining of ChIFNγ

Cultured cells were resuspended in U bottom 96-well plates in FACS buffer (PBS containing 1.0% BSA and 0.1% sodium azide) and incubated with normal mouse serum (1%) for 10 min at RT to block non-specific binding. For surface staining, cells were incubated, protected from light, for 15 min at room temperature with appropriate dilutions of antibodies: mouse anti-chicken CD4-FITC (CT4) (Southern Biotech), CD8α-PerCp (Tregaskes et al., 1995). After incubation, cells were washed twice with FACS buffer and were either used for intracellular staining or fixed with a solution of 2% paraformaldehyde in PBS.

Incubation with primary antibodies to MHC I (Salomonsen et al., 1987) and MHC II (Kaufman et al., 1990) was followed by Alexa-647 conjugated goat anti-mouse antibody (Life Technologies). Secondary antibody alone or unconjugated goat anti-mouse antibody (Life Technologies) was used as an unstained control for surface MHC staining. Intracellular staining was carried out as described previously (Ariaans et al., 2008). Briefly, splenocytes from challenged birds or non-infected controls were seeded in a 96-well round-bottom plates (Nunc) at $10^6$ cells/well in a final volume of 200 μl of culture media or culture media supplemented with the different stimuli at the concentration described in the ELISpot technique (except PMA which was used at 50 ng/ml). Cells were cultured using the conditions described above for ELISpot assays (24 h culture). For intracellular staining, during the last 2 h of culture, cells were treated with Brefeldin A according to the manufacturer's instructions (Cytofix/Cytoperm™ Plus Fixation/Permeabilization kit, BD Biosciences). To avoid non-specific binding signal, we preincubated cells with Cytofix/Cytoperm™ buffer containing 2% normal mouse serum and further staining steps involved Cytofix/Cytoperm™ washing buffer containing 1% normal mouse serum (Biosource). To confirm the specificity of the anti-IFNγ antibody EH9 we also employed a validated anti-IFNγ [mAb80 (Ariaans et al., 2008)]. Purified fractions of both antibodies were conjugated using Alexa Fluor® 647 monoclonal antibody labeling kit (Molecular Probes) according to the manufacturer's instructions. A mouse isotype matched control antibody IgG1 Alexa Fluor® 647 (Life Technologies) was employed at the same concentration as EH9 and Mab80. For analysis, a gate on the FSC/SSC region of lymphocytes was selected and a minimum of 10,000 events were acquired on a FACSCalibur instrument using Cell Quest software (BD Becton Dickinson). Flow cytometry analysis indicates that non-adherent CKC were not present at significant levels (data not shown). FlowJo software (TreeStar) was used to analyze flow cytometry data.

## 2.11. Statistical analysis

A paired or unpaired t-student test or one-way ANOVA was performed using GraphPad Prism (version 6.0 for Windows, GraphPad Software, San Diego, California, USA).

# 3. RESULTS

## 3.1. Validation of anti-chicken interferon gamma antibodies

Screening identified two anti-chicken IFNγ antibodies (clones EH9 and AF10) which were shown by ELISA to bind recombinant chicken IFNγ and to work effectively as an antibody pair in capture ELISA (Supplementary Fig. 2A–C). We subsequently compared this antibody pair with commercially available antibodies [from Life Technologies (Ariaans et al., 2008 and Reemers et al., 2012)] in ELISpot assays. There were no differences in the numbers of spots from the same samples of chicken splenocytes stimulated with Con A using either antibody pairs (EH9/AF10: $338.1 \pm 23.2$ SFU/$10^6$cells; Life Technologies: $338.9 \pm 21.3$ SFU/$10^6$ cells, $n = 9$ (Supplementary Fig. 2D)).

## 3.2. Viral shedding dynamics and immune responses

Following intranasal infection of Line O birds with LPAI H7N7, buccal swab samples were analyzed for the presence of influenza M1 transcript by qRT-PCR. These were found to be positive from the earliest sampling time point at day 4 post-infection. Viral transcript was still detectable albeit at a lower level ($p < 0.05$) in the buccal swabs at day 6, and was undetectable at day 10 (data not shown). Challenged birds exhibited significantly higher HI titers compared to non-infected controls (Fig. 1A, $p < 0.01$). All subsequent experiments were performed in Line O birds, with the exception of the vaccine cohort (Line 15, final figure).

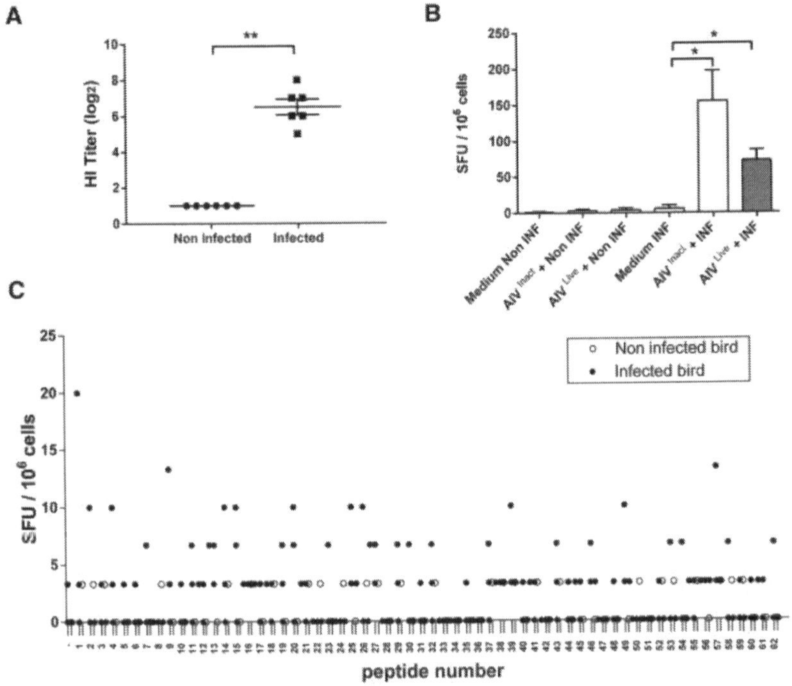


**Figure. 1.** Viral shedding, serum antibody titers and ELISpot responses of challenged birds. Birds were challenged with LPAI influenza strain H7N7/turkey/Eng/77. A) Influenza antiserum titers, HI assay from non-infected birds and infected birds; results expressed as mean ± S.E.M. (n = 6 per group). B) ChIFNγ ELISpot responses of splenocytes from infected birds (INF, n = 4) sacrificed 2 weeks after challenge and non-infected birds (non-INF, n = 3) were cultured with neutralized influenza virus ($AIV^{Inact}$) and live virus ($AIV^{Live}$). Results expressed as mean (± S.E.M.) of spot forming units (SFU) per million cells. Significance of data is represented as follows: * = $p < 0.05$, ** = $p < 0.01$. SFU: spot forming units. C) ChIFNγ ELISpot responses of peptide library against influenza viral nucleoprotein (NP). In total, 62 different peptides were analyzed in splenocytes from individual birds infected at 2 weeks post-challenge (black dot, n = 3) and control not infected (white circle, n = 1).

We tested our antibody pair for use in ELISpot with live or beta propiolactone inactivated challenge-strain virus to stimulate splenocyte responses (Fig. 1B). Splenocytes from control (non-infected) birds did not produce IFNγ when exposed to either live or inactivated virus. In contrast, splenocytes from infected birds did produce IFNγ ($p < 0.05$) following exposure to both live ($72.0 \pm 15.4$ SFU/$10^6$ cells) and inactivated virus ($155.2 \pm 42.3$ SFU/$10^6$ cells), as expected. The use of live virus consistently yielded lower responses than the use of inactivated virus in all samples, although this difference was not statistically significant.

To identify epitope-specific responses, we employed an NP peptide library corresponding to the challenge virus. We analyzed responses to pooled peptides at 1 week post-infection (Supplementary Fig. 3) and to individual peptides 2 weeks after infection (Fig. 1C). Responses to individual peptides were low, not consistent between birds, and not statistically significantly different between control and infected birds. In the following experiments, an alternative strategy to detect specific IFN responses was developed.

## 3.3. CKCs express only MHC class I following repeated passage

To potentiate the detection of influenza-specific CD8 T cell responses, we generated a CKC cell line expressing only MHC class I. We passaged CKC from Line O birds a minimum of eight times. We then analyzed the cells by flow cytometry for the expression of MHC classes I and II. The passaged CKC were found to exclusively express MHC class I (Fig. 2).

## 3.4. ELISpot using infected CKC co-culture method

Having validated the necessary individual components we introduced the method of co-culture of responding cells with infected CKCs. Despite the fact that so many antigen specific cells were detectable in co-culture with infected CKCs, the background response for this assay was extremely low (control and INFγ only data, Fig. 3), demonstrating its specificity and sensitivity. Splenocytes from infected birds (2 weeks post-infection) produced extremely high (mean: $833 \pm 134$ SFU/$10^6$ cells) numbers of spot forming units when co-cultured with infected CKC (Fig. 3). This response was significantly different ($p < 0.01$) from the response seen with infected CKC co-cultured with splenocytes from non-infected (control) birds. The SFU count seen with co-culture of infected CKC with infected splenocytes was close to that seen with cells from infected birds stimulated with PMA/ionomycin (1060 ± 53 SPU/$10^6$ cells), suggesting that antigen specific antiviral IFNγ producing cells constitute the majority of those able to rapidly produce IFNγ. It was interesting to note that splenocytes from infected birds have greater SFU responses to PMA in our study (discussed below).

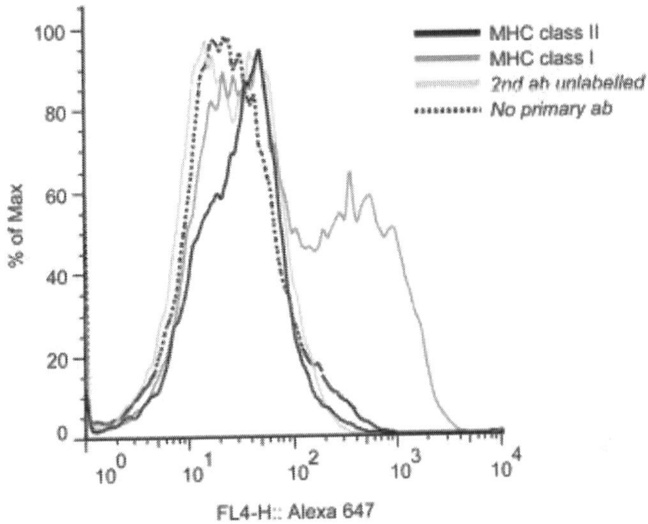


**Figure. 2.** Analysis of MHC class I and II expression on CKC passaged at least 8 times. Expression of MHC classes I and II on CKC detected using an Alexa 647 labeled secondary antibody. Dark line = class II, dark gray line = class I, light gray line = unlabelled secondary antibody, dashed line = only Alexa 647 labeled secondary antibody.

## 3.5. Analysis of IFNγ positive cells from co-culture assay

To analyze the phenotype of the responding splenocytes from infected birds we performed intracellular staining on cells from co-culture assays. We first validated antibody (EH9) against a previously published anti

IFNγ antibody (mAb80, (Ariaans et al., 2008)) using IFNγ transfected CHO cell lines (Supplementary Fig. 4) and in splenocytes stimulated with PMA/ionomycin (Fig. 4A). There was no statistically significant difference between results obtained with the two antibodies. Non-specific signal was not detected by isotype control staining (Fig. 4B).

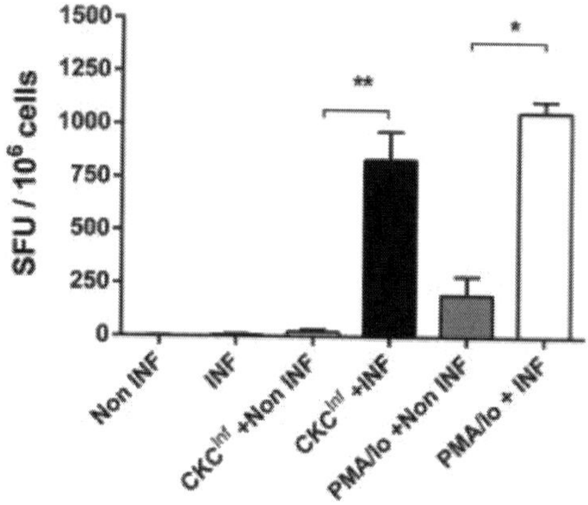


**Figure. 3.** Splenocyte IFNγ co-culture ELISpot responses to infected CKC. Splenocytes from infected birds (INF, n = 4) sacrificed 2 weeks after challenge and control non-infected birds (non-INF, n = 3) were cultured with infected chicken kidney cells ($CKC^{INF}$). Results expressed as mean (± S.E.M.) of spot forming units (SFU) per million cells. Significance of the data is represented as follows: * = $p < 0.05$, ** = $p < 0.01$. SFU: spot forming units.

We then analyzed the phenotype of IFNγ expressing cells from infected birds, following co-culture with either infected or non-infected CKC. Data shown are for a representative sample from infected and non-infected birds (Fig. 4C) gating in the same FSC/SSC lymphocyte region (Fig. 4A) for all conditions. The greatest number of interferon gamma producing cells was detected during co-culture of infected CKC with splenocytes from infected birds (0.517%), compared with splenocytes from infected birds co-cultured with non-infected CKC (0.069%), and splenocytes from non-infected birds co-cultured with infected CKC (0.071%). It is important to note that the majority of IFNγ positive splenocytes from infected birds co-cultured with infected CKC were CD8 positive (> 60%,Fig. 4C).

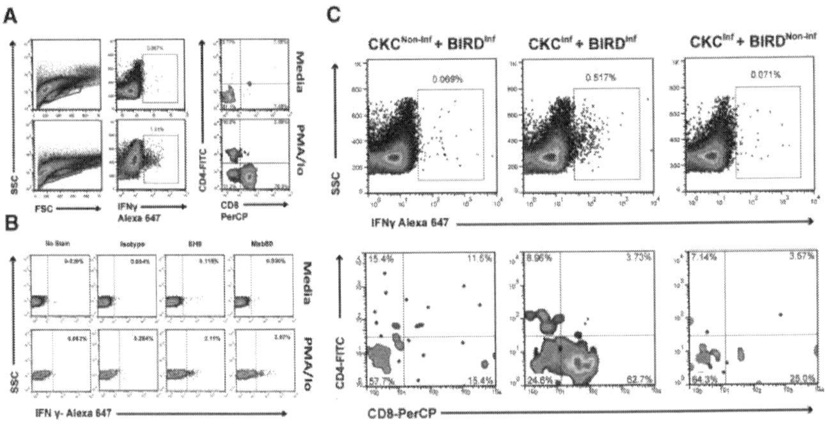


**Figure. 4.** Intracellular IFNγ staining and phenotype of responding lymphocytes following co-culture. A) Representative example of gating of live lymphocytes in intracellular staining. Lymphocyte gate was designated for the region of lymphocytes stimulated with PMA/Io (bottom left figure) and applied to other culture conditions (media alone, upper left figure). B) Comparison of EH9 (in house) and mAb80 (previously described) intracellular staining for IFNγ; both antibodies detected a similar percentage of $IFN\gamma^+$ cells in media control or PMA/ionomycin stimulated splenocytes. Matched IgG isotype control was also employed. C) Representative example of intracellular staining of CKC co-cultured with splenocytes from an infected bird ($BIRD^{INF}$) or control non-infected bird ($BIRD^{NON\text{-}INF}$): upper row figures, percentage of IFNγ positive cells following co-culture with non-infected ($CKC^{non\text{-}Inf}$) or infected ($CKC^{Inf}$) CKC; lower row figures, $CD4^+$and $CD8^+$ lymphocyte populations in the IFNγ positive populations from above figures.

## 3.6. Detection of antigen specific IFNγ responses in splenocytes from recombinant Fowlpox vaccinated birds employing CKC infected with recombinant virus MVAGFP/MVANpM1 after influenza challenge

Having established the utility of the co-culture ELISpot we used the technique to analyze influenza antigen specific responses in birds vaccinated (prime and boost) with recombinant Fowlpox (F9) or recombinant Fowlpox-NpM1 (F9-NpM1), and then challenged with an influenza virus with heterologous nucleoprotein and matrix protein. Instead of infecting the CKC with influenza virus we used recombinant MVA carrying either a GFP or NpM1 fusion transgene (homologous to the Fowlpox recombinant) then irradiated the infected CKC as described.

Three of the four F9-NpM1 vaccinated birds challenged with influenza showed IFNγ responses that distinguished them from F9 vaccinated and challenged birds (Fig. 5) ($40.0 \pm 12.5$ vs. $3.0 \pm 1.9$, $p < 0.05$). The majority of responses in the F9-NpM1 vaccinated birds were greater with CKC infected with MVA-NpM1 fusion transgene. Some responses were also observed with F9-NpM1 vaccinated birds when APCs were infected with MVA-GFP (although this result was not significant). Our results with vaccinated birds thus suggest that this technique can be used to evaluate vaccine induced responses, and additionally that recombinant viruses expressing appropriate transgenes can be used to replace influenza infection of CKC. This permits the analysis of more defined antigen specific responses while reducing the requirement to handle live influenza virus in the laboratory.

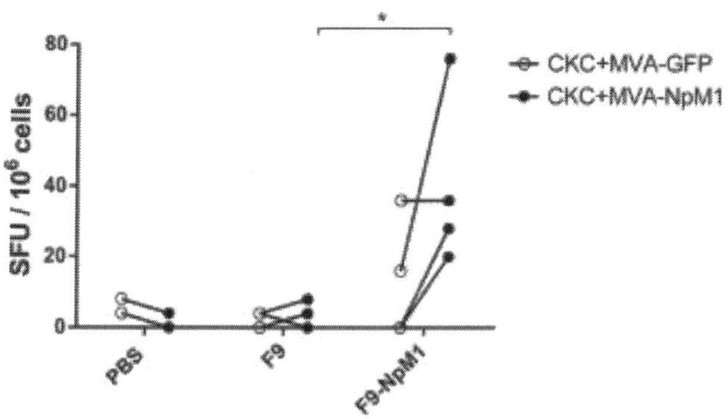


**Figure. 5.** IFNγ co-culture ELISpot using CKC infected with recombinant MVA and splenocytes from birds challenged with LPAI. ELISpot responses in splenocytes from infected birds from unvaccinated (PBS group, n = 3) or birds vaccinated with a recombinant vaccine Fowlpox (F9 group, n = 4), or Fowlpox$^{NpM1}$ (F9-NpM1 group, n = 4). Splenocytes were cultured with either irradiated MVA$^{GFP}$ infected CKC (black dot) or irradiated MVA$^{NpM1}$ (white circle) infected CKC. Lines indicate samples from the same spleen under the two previous conditions. Results were expressed as spot forming units (SFU) per million cells. Significance of the data is represented as follows: * = $p < 0.05$.

# 4. DISCUSSION

We have developed a method to potentiate the detection and analysis of influenza antigen specific T cells utilizing infected CKC to present viral peptides in a manner biologically relevant to CD8 T cells. We have demonstrated that our co-culture ELISpot detects greater numbers of antigen specific CD8 T cells than ELISpot with whole virus as an antigen. Our assay can also be adapted to use recombinant viruses to infect CKC, increasing its specificity and reducing the requirement to work with live influenza virus. Our results are the first to demonstrate detection by flow cytometry of influenza-specific IFNγ responses in individual T cells from LPAI infected birds.

The ability of our method to detect such large numbers of antigen specific T cells (similar numbers to positive controls with PMA/ionomycin, see example Supplementary Fig. 5) likely reflects not only the high promiscuity of the B21 haplotype, but also the fact that our CKC cell line expresses only MHC class I and presents peptides following a biologically relevant infection process.

In ELISpot using whole influenza virus we were able to detect antigen specific responses, although these were much lower (Fig. 1). Although ELISpot has previously been used to measure antiviral responses against other avian viruses, including NDV (Ariaans et al., 2008) and IBV (Ariaans et al., 2009), it has never been employed to analyze avian responses to influenza. In the present study, following challenge with H7N7 LPAI, the birds became serologically positive and showed specific IFNγ responses, irrespective of whether inactivated or live avian influenza virus was added to endogenous APCs (Fig. 1). Additionally, ELISpot with live virus added to splenocytes from infected birds further reduced SFU counts. It is possible that live virus affects the interactions, and/or the functionality, of cells in vitro (Hinshaw et al., 1994, Oh et al., 2000 and Hao et al., 2008).

It was interesting to note that splenocytes from infected birds have greater SFU responses to PMA in our study. PMA does not activate all T cells (Suzawa et al., 1984 and Kim et al., 1986)., It may be that antigen experienced cells (from infected birds) have a lower threshold of activation and are activated more readily by PMA, hence the higher SFU counts in the infected cohort positive control compared with the non-infected. Another possibility is altered lymphocyte subset frequencies in infected birds. We did not note any difference in the relative proportions of B and T lymphocytes between infected and non-infected birds (data not shown) but as detailed analysis of T cell subsets was not performed we cannot discount a greater frequency of a subset more responsive to PMA in the infected birds. Given the magnitude of the difference we consider this second possibility less likely. Unfortunately given the paucity of this type of data in this area in avian immunology we have not been able to make extensive direct comparisons, other than to observe that our positive control results are in the range reported by the few directly comparable studies of ELISpot and/or intracellular staining (Ariaans et al., 2008 and Ariaans et al., 2009); however these do not report directly comparable infection data. In the only study regarding the phenotype of responding cells during HPAI infection of chickens (Seo et al, 2002), employing different methods, the percentage of IFNγ producing CD8 positive cells in the spleen was approximately 50% at day 6 post-infection, falling to an average of 15% at 20 days post-infection. This result is much higher than that detected in infected birds in our study; however Seo et al. did not distinguish between IFNγ producing T cells and IFNγ from NK cells, which may account for the difference. We could detect no evidence for NK activation using our method as we were not able to detect a significant number of IFNγ positive cells with splenocytes from non-infected birds cultured with infected CKC (Fig. 4C), or with splenocytes from infected birds cultured with non-infected CKCs (Supplementary Fig. 5). While our study did not identify the TCR subtype of the IFNγ producing CD8 positive cells, it has been hypothesized that the main population involved in IFNγ responses and in viral clearance is TCR αβ (Vβ1, TCR2) (Seo et al., 2002). Interestingly, the control of acute IBV infection has also been attributed to CD8-TCR2 lymphocytes (Collisson et al., 2000). Further studies are required to identify the TCR subsets responsible for the immune response in our model.

Our co-culture method was better able to distinguish responses between infected and control birds than ELISpot using a peptide library. In comparison with recently published work using a high concentration of peptides to analyze influenza-specific responses (Reemers et al., 2012), the co-culture ELISpot is more sensitive and has a significantly lower background. However unlike peptide assays, it lacks precise epitope specificity and cannot distinguish responses against individual proteins. We demonstrated a further level of specificity by infecting CKC with an MVA recombinant virus expressing a fusion protein (NpM1) from a human H3N2 virus (Berthoud et al., 2011). These cells were used to present antigens to splenocytes from birds given a recombinant Fowlpox vaccine, also expressing nucleoprotein and matrix protein 1, and then challenged with a heterologous LPAI virus. Although the NpM1 sequences of the MVA, Fowlpox recombinants and challenge virus were not homologous, these are highly conserved (Lillie et al., 2012) internal influenza antigens (example 98% homology for NP and 100% for M1 protein, Supplementary Fig. 6). We have already shown cross protective immune responses using the MVA vector and the same strain of influenza virus (Boyd et al., 2012). In the current study, we were able to distinguish if individual infected birds were vaccinated or not, since the vaccinated group possessed higher specific responses than unvaccinated birds. Our results suggest that infection of CKC with recombinant virus containing transgenes for an epitope of interest could be used to increase the sensitivity of assays to detect antigen and epitope specific T cells.

In summary we have developed a sensitive method for the detection of antigen specific T cells, which will be important in the analysis of immune responses to both vaccines and pathogens. The assay provides greater sensitivity than the use of inactivated or live virus in ELISpot, and reduced background compared with peptide library ELISpot. Our method is also more accessible to a wider community than methods employing expensive peptide libraries, the interpretation of which data is rendered problematic due to an incomplete knowledge of avian MHC binding specificities. While we have demonstrated its efficacy for influenza, this technique can be applied to the study of T cell responses for many avian pathogens. We also demonstrated that the use of recombinant virus to infect CKC can further define antigen specificity, and additionally reduce the requirement to handle live zoonotic pathogen, an important safety consideration.

## ACKNOWLEDGMENTS

We thank Dr. Mike Skinner for providing the Fowlpox construct and Dr. Sarah Gilbert for providing the MVA constructs. We thank John R. Young for his comments and help in analyzing data. We would like to acknowledge the expert and dedicated help of the Animal and Media Services staff at the Pirbright Institute. This work was funded by the BBSRC (Biotechnology and Biological Sciences Research Council) under grant numberBB/E011691/1. The funders had no role in study design, data collection and analysis, decision to publish, or preparation of the manuscript.

## REFERENCES

1 Ariaans, M.P., van de Haar, P.M., Lowenthal, J.W., van Eden, W., Hensen, E.J., Vervelde, L., 2008. ELISPOT and intracellular cytokine staining: novel assays for quantifying T cell responses in the chicken. Dev. Comp. Immunol. 32, 1398.

2 Ariaans, M.P., van de Haar, P.M., Hensen, E.J., Vervelde, L., 2009. Infectious Bronchitis Virus induces acute interferon-gamma production through polyclonal stimulation of chicken leukocytes. Virology 385, 68.

3 Berthoud, T.K., Hamill,M., Lillie, P.J., Hwenda, L., Collins, K.A., Ewer, K.J., Milicic, A., Poyntz, H.C., Lambe, T., Fletcher, H.A., Hill, A.V., Gilbert, S.C., 2011. Potent CD8+ T-cell immunogenicity in humans of a novel heterosubtypic influenza A vaccine, MVA-NP + M1. Clin. Infect. Dis. 52, 1.

4 Boyd, A.C., Ruiz-Hernandez, R., Peroval, M.Y., Carson, C., Balkissoon, D., Staines, K., Turner, A.V., Hill, A.V., Gilbert, S.C., Butter, C., 2012. Towards a universal vaccine for avian influenza: protective efficacy of modified Vaccinia virus Ankara and Adenovirus vaccines expressing conserved influenza antigens in chickens challenged with low pathogenic avian influenza virus. Vaccine 31, 670.

5 Collisson, E.W., Pei, J., Dzielawa, J., Seo, S.H., 2000. Cytotoxic T lymphocytes are critical in the control of infectious bronchitis virus in poultry. Dev. Comp. Immunol. 24, 187.

6 Dalgaard, T.S., Norup, L.R., Rubbenstroth, D., Wattrang, E., Juul-Madsen, H.R., 2010. Flow cytometric assessment of antigen-specific proliferation in peripheral chicken T cells by CFSE dilution. Vet. Immunol. Immunopathol. 138, 85.

7 De Boever, S., Croubels, S., Demeyere, K., Lambrecht, B., De Backer, P., Meyer, E., 2010. Flow cytometric differentiation of avian leukocytes and analysis of their intracellular cytokine expression. Avian Pathol. 39, 41.

8 Gaush, C.R., Smith, T.F., 1968. Replication and plaque assay of influenza virus in an established line of canine kidney cells. Appl. Microbiol. 16, 588.

9 Gobel, T.W., Schneider, K., Schaerer, B., Mejri, I., Puehler, F., Weigend, S., Staeheli, P., Kaspers, B., 2003. IL-18 stimulates the proliferation and IFNgamma release of CD4+ T cells in the chicken: conservation of a Th1-like system in a nonmammalian species. J. Immunol. 171, 1809.

10 Haghighi, H.R., Read, L.R., Haeryfar, S.M., Behboudi, S., Sharif, S., 2009. Identification of a dual-specific T cell epitope of the haemagglutinin antigen of an h5 avian influenza virus in chickens. PLoS ONE 4, e7772.

11 Hao, X., Kim, T.S., Braciale, T.J., 2008. Differential response of respiratory dendritic cell subsets to influenza virus infection. J. Virol. 82, 4908.

12 Hinshaw, V.S., Olsen, C.W., Dybdahl-Sissoko, N., Evans, D., 1994. Apoptosis: a mechanism of cell killing by influenza A and B viruses. J. Virol. 68, 3667.

13 Home Office, U., 1986. (1986) Guidance on the Operation of the Animals (Scientific Procedures) Act.

14 Jonges, M., Liu,W.M., van der Vries, E., Jacobi, R., Pronk, I., Boog, C., Koopmans, M., Meijer, A., Soethout, E., 2010. Influenza virus

inactivation for studies of antigenicity and phenotypic neuraminidase inhibitor resistance profiling.

15 J. Clin. Microbiol. 48, 928.

16 Kaufman, J., Skjoedt, K., Salomonsen, J., Simonsen,M., Dupasquier, L., Parisot, R., Riegert, P., 1990. MHC-like molecules in some nonmammalian vertebrates can be detected by some cross-reactive xenoantisera. J. Immunol. 144, 2258.

17 Kim, D.K., Otten, G., Moldwin, R.L., Dunn, D.E., Nau, G.J., Fitch, F.W., 1986. PMA alone induces proliferation of somemurine T-cell clones but not others. J. Immunol. 137, 2755.

18 Koch,M., Camp, S., Collen, T., Avila, D., Salomonsen, J.,Wallny, H.J., van Hateren, A., Hunt, L., Jacob, J.P., Johnston, F., Marston, D.A., Shaw, I., Dunbar, P.R., Cerundolo, V., Jones, E.Y., Kaufman, J., 2007. Structures of an MHC class I molecule from B21 chickens illustrate promiscuous peptide binding. Immunity 27, 885.

19 Kwon, J.S., Lee, H.J., Lee, D.H., Lee, Y.J.,Mo, I.P.,Nahm, S.S., Kim,M.J., Lee, J.B., Park, S.Y., Choi, I.S., Song, C.S., 2008. Immune responses and pathogenesis in immunocompromised chickens in response to infectionwith theH9N2 low pathogenic avian influenza virus. Virus Res. 133, 187.

20 Lillie, P.J., Berthoud, T.K., Powell, T.J., Lambe, T., Mullarkey, C., Spencer, A.J., Hamill, M., Peng, Y., Blais,M.E.,Duncan, C.J., Sheehy, S.H.,Havelock, T., Faust, S.N., Williams, R.L., Gilbert, A., Oxford, J., Dong, T., Hill, A.V., Gilbert, S.C., 2012. Preliminary assessment of the efficacy of a T-cell-based influenza vaccine, MVA-NP + M1, in humans. Clin. Infect. Dis. 55, 19.

21 Miller,M.M., Bacon, L.D., Hala, K., Hunt, H.D., Ewald, S.J., Kaufman, J., Zoorob, R., Briles,W.E., 2004. 2004 nomenclature for the chicken major histocompatibility (B and Y) complex. Immunogenetics 56, 261.

22 Novak, R., Ester, K., Savic, V., Sekellick,M.J., Marcus, P.I., Lowenthal, J.W., Vainio, O., Ragland, W.L., 2001. Immune status

assessment by abundance of IFNalpha and IFN-gamma mRNA in chicken blood. J. Interferon Cytokine Res. 21, 643.

23 Oh, S., McCaffery, J.M., Eichelberger, M.C., 2000. Dose-dependent changes in influenza virus-infected dendritic cells result in increased allogeneic T-cell proliferation at low, but not high, doses of virus. J. Virol. 74, 5460.

24 Rauw, F., Anbari, S., van den Berg, T., Lambrecht, B., 2011. Measurement of systemic and local respiratory cell-mediated immunity after influenza infection in chickens. Vet. Immunol. Immunopathol. 143, 27.

25 Reemers, S.S., van Haarlem, D.A., Sijts, A.J., Vervelde, L., Jansen, C.A., 2012. Identification of novel avian influenza virus derived CD8+ T-cell epitopes. PLoS ONE 7, e31953.

26 Rock, K.L., Gamble, S., Rothstein, L., 1990. Presentation of exogenous antigen with class I major histocompatibility complex molecules. Science 249, 918. Salomonsen, J., Skjodt, K., Crone, M., Simonsen, M., 1987. The chicken erythrocyte-specific MHC antigen — characterization and purification of the B-G antigen by monoclonal-antibodies. Immunogenetics 25, 373.

27 Seo, S.H., Collisson, E.W., 1997. Specific cytotoxic T lymphocytes are involved in in vivo clearance of infectious bronchitis virus. J. Virol. 71, 5173.

28 Seo, S.H., Peiris, M., Webster, R.G., 2002. Protective cross-reactive cellular immunity to lethal A/Goose/Guangdong/1/96-like H5N1 influenza virus is correlated with the proportion of pulmonary CD8(+) T cells expressing gamma interferon. J. Virol. 76, 4886.

29 Singh, S., Briles, W.E., Lupiani, B., Collisson, E.W., 2010a. Avian influenza viral nucleocapsid and haemagglutinin proteins induce chicken CD8+memory T lymphocytes. Virology 399, 231.

30 Singh, S., Toro, H., Tang, D.C., Briles, W.E., Yates, L.M., Kopulos, R.T., Collisson, E.W., 2010b. Non-replicating adenovirus vectors expressing avian influenza virus haemagglutinin and nucleocapsid

proteins induce chicken specific effector, memory and effector memory CD8(+) T lymphocytes. Virology 405, 62.

31 Spackman, E., 2008. Avian Influenza Virus. Humana. Springer, distributor, London, Totowa, N.J.

32 Spackman, E., Senne, D.A., Myers, T.J., Bulaga, L.L., Garber, L.P., Perdue, M.L., Lohman, K., Daum, L.T., Suarez, D.L., 2002. Development of a real-time reverse transcriptase PCR assay for type A influenza virus and the avian H5 and H7 haemagglutinin subtypes. J. Clin. Microbiol. 40, 3256.

33 Staines, K., Young, J.R., Butter, C., 2013. Expression of chicken DEC205 reflects the unique structure and function of the avian immune system. PLoSONE 8.

34 Stech, O., Veits, J., Weber, S., Deckers, D., Schroer, D., Vahlenkamp, T.W., Breithaupt, A., Teifke, J., Mettenleiter, T.C., Stech, J., 2009. Acquisition of a polybasic haemagglutinin cleavage site by a low-pathogenic avian influenza virus is not sufficient for immediate transformation into a highly pathogenic strain. J. Virol. 83, 5864.

35 Suarez, D.L., Schultz-Cherry, S., 2000. Immunology of avian influenza virus: a review. Dev. Comp. Immunol. 24, 269.

36 Subbarao, K., Joseph, T., 2007. Scientific barriers to developing vaccines against avian influenza viruses. Nat. Rev. Immunol. 7, 267.

37 Suzawa, T., Hyodo, S., Kishi, T., Kashiwa, H., Karakawa, T., Kittaka, E., Sakano, T., Usui, T., 1984. Comparison of phorbol-myristate acetate and phytohemagglutinin as stimulators of in vitro T lymphocyte colony formation of human peripheral-blood lymphocytes. 1. Surface-markers of colony cells. Immunology 53, 499.

38 Tregaskes, C.A., Kong, F.K., Paramithiotis, E., Chen, C.L.H., Ratcliffe,M.J.H., Davison, T.F., Young, J.R., 1995. Identification and analysis of the expression of CD8- alpha-beta and CD8-alpha-alpha isoforms in chickens reveals a major TCRgamma–delta CD8-

alpha–beta subset of intestinal intraepithelial lymphocytes. J. Immunol. 154, 4485.

39 World Health Organization. Dept. of Epidemic Response Pandemic Alert and response, 2002. WHO Manual on Animal Influenza Diagnosis and Surveillance. World Health Organization, Geneva.

# INDEX